SQUATCH COP

INVESTIGATING AND DOCUMENTING THE BIGFOOT PHENOMENON

By
Larry Baxter

www.alasquatchpodcast.com

SQUATCH COP
Investigating and Documenting the Bigfoot Phenomenon

ISBN: 978-1-7367297-2-4 (Paperback)
ISBN: 978-1-7367297-3-1 (E-book)

Library of Congress Control Number: 2022911639

Published by Larry A. Baxter
Edited by Tammaron K. Baxter

Cover Photo by Larry A. Baxter
Cover Design by Tammaron K Baxter

Alasquatch Publishing
PO Box 887
Homer, Alaska 99603

www.alasquatchpodcast.com

Also by Larry Baxter

Abandoned: The History and Horror of Port Chatham, Alaska

Dedication

This book is dedicated to the men and women of law enforcement who have stepped (or leapt) out of their comfort zones to investigate the unknown.

CHAPTER 1
WHO LET YOU IN HERE?

I started my law enforcement career in Alaska in 2006. I took a job on a small Pribilof Island in the middle of the Bering Sea. I wanted to be a cop so bad that I was willing to do it anywhere. I got a lot of experience on that tiny island. Working everything from traffic accidents to homicide.

In those early years of my career, I found I had a knack and proclivity for investigations. I often gravitated toward complex or more involved cases that other officers would shy away from. I moved from that tiny island to a larger (but not by much) city and eventually found myself promoted to Investigations Sergeant. I was conducting or supervising all significant investigations, including murder, sexual assault, financial crimes, and other felonious activities.

My new department was busier, but they were also generous with training. As a result, I got to attend a lot of training. I've had training in interview and interrogation, child and adult homicide investigation, evidence collection and management, digital forensics, sexual assault investigation, active shooter situations, and many other disciplines. Without exaggeration, I can confidently say I have had hundreds of hours of training.

Many people do not realize that small departments can often forge better, more well-rounded officers. For example, in a large city like New York or Los Angeles, a patrol officer responding to homicide or other major crime would, most likely, be relegated to guarding the scene, performing crowd control, or other menial tasks. Meanwhile, detectives and the crime scene technicians would process the scene, collect evidence and interview witnesses, suspects, and victims. However, in a smaller department, there may not be a stable of detectives or a van full of crime scene technicians to work the scene. Often, small-town cops get to be the initial responders, crime scene technicians, and detectives, all in one.

I'm not telling you this to brag, but I do want you to know what qualifies me to sit here and write on this subject. I've always been interested in things unexplained like Bigfoot, UFOs, and ghosts, but I've always gravitated toward Bigfoot when it comes to investigating strange things.

I always felt there was plenty of information on how to look for Bigfoot, but there is a lack of material for citizen scientists on how to investigate Bigfoot. Of course, it's great to get out in the field and speak with witnesses, cast tracks, and take photographs, but what do you do with that stuff, and how do you organize it?

I wish I could tell you that if you follow my advice, you will be successful in proving the existence of Bigfoot. However, given the amount of time and effort that has already gone into the pursuit of the elusive cryptid, I doubt that anyone in our lifetime will ever get definitive answers to the mystery of Bigfoot. But, like the man says, "It's not about the destination, it's the journey." Hopefully, you will find some valuable tools in the toolbox I'm about to open for you. Take what you can use and put it to good use. Maybe, just maybe, it will help us to solve one of the greatest mysteries ever.

CHAPTER 2
SO, YOU WANT TO LOOK FOR BIGFOOT?

Despite what you see on television, no one is born a police officer. At one time, every officer was once a rookie and had to go through training. I do not know of any state in the United States that does not require the attendance of some kind of academy or training program for new officers.

The same cannot be said for those of us with the rare hobby of looking for Bigfoot. Some organizations offer paid expeditions that Bigfoot enthusiasts can participate in, which offer some guidance and training in collecting and preserving evidence. However, time and space are limited on these expeditions, and the cost can be somewhat prohibitive to those on a budget.

One of my favorite things about Bigfoot research is all the different aspects that it entails. When most people think of Bigfoot research and investigation, they conjure up mental images of people tromping through the woods, belting out whoops and screams, and beating trees with baseball bats. Well, sometimes, that does happen. However, there are many other ways to conduct research. For example, an armchair researcher who peruses historical newspapers online has the potential to discover information that could collaborate behavior reported by a witness or refute a theory put forth by another researcher. There are many ways that someone can contribute to the search for Bigfoot, and not all of them require going into the woods.

Much of this book will focus on fieldwork and evidence collection, preservation, and documentation of evidence found in the field. However, even if you are not a field researcher, I suggest you stick with us and persevere. I will discuss evidence collection and organizing and storing evidence, not just physical evidence but also digital evidence and evidence found online.

CHAPTER 3
EQUIPMENT AND GEAR

One of the best moments of a new officer's career is when they get to put on their duty belt for the first time. Being a fan of comic books for many years before becoming a law enforcement officer, I liked to think of my duty belt as my "utility belt." Later in my career, I likened it more to a torture device that made my back sore and forever changed the way I walk (I still walk like I'm wearing it sometimes).

While attending the police academy, we actually had a class on arranging the gear on the duty belt and how to wear it. At the end of my career, I found most of the equipment on my belt was still in the same place it was after that class. They instructed us on how to arrange the gear to lessen our chance of back injuries, and they also gave us some essential advice regarding gear placement. I didn't realize it at the time, but many of the lessons and experience I gained in my law enforcement career, would actually transfer over, very appropriately, to Bigfooting.

Some of the best advice I have ever received regarding gear and equipment is to be consistent! Once you find out what arrangement works for you, keep your gear like that and wear it the same way every time. In time, your body becomes accustomed to specific movements and actions. For example, suppose you carry your camera in your right shirt pocket every day. When you need your camera, you will automatically reach for your right shirt pocket without even thinking about it. You will have trained your muscles to perform the action of going for your camera without even consciously putting any thought into it.

Muscle memory is a genuine phenomenon, and I imagine most of us have experienced it on one level or another. It simply consolidates a specific task (like reaching for your phone, gun, or camera). When working as a Field Training Officer, I would instruct my new officers to experiment with their gear placement and find what worked. After you find your happy place, stay there. I encouraged them to keep those items in place and not move them around once they found a good position on their belt for their handcuffs, pepper spray, and taser.

I know from experience; that it can be very distracting to reach for a piece of gear, and it's not there. For example, imagine trying to control a resisting suspect, and you reach down for your handcuffs, and they are not there! That feeling can distract you from the moment and put yourself and your partners in danger. That's why it is essential to be consistent with your gear placement, so you can be confident that it will be there when you reach for it.

Now, take this same concept and transfer it over to Bigfooting. Experiment with your gear and find a loadout that works for you. For example, you may like to keep your camera around your neck or in your cargo pants pocket. Find a comfortable and convenient place and keep it there every time you go out in the field. Practice deploying it quickly, and find a way that works for you.

One of the challenges of gear placement is prioritization. Getting the gear you need most, where you can get it the fastest. Many factors can go into the decision of where to keep your equipment, depending on your goals. If your ultimate goal is to get video or photos of Bigfoot, your camera equipment will take priority. You will want to set your gear up to get to your camera as quickly as possible. Further, you will want to keep the equipment you need to keep your camera up and running, like batteries, lens wipes, and memory cards, readily accessible and able to deploy.

It would be disappointing to have a Sasquatch in your camera lens, only to discover the battery is dead and the fresh batteries are in the backpack on your back, deep in a bundle of clothes and gear. Think about gear and equipment you might need "right now!" and try to position that gear within arm's reach, like keeping at least one spare camera battery in your front jeans pocket or another similar setup. Mentally run through scenarios and prepare for different problems that may arise. You can also learn a lot from others' mistakes. Listen to other enthusiasts who have shared their tales of mistakes and failures and compare them with how you operate. Remember, sometimes even a bad example can be helpful if it teaches you how to do something better.

This philosophy applies to the gear on your person and the equipment in your pack. Once you find a pack that works for you, be consistent in where you store your gear in the bag. Always keep the batteries, tape measurer, etc., in the same place to know where they are. This will save you time in the long run. I have seen and been the person looking for a small item and having to completely empty a backpack to find it.

This can be applied to your vehicle as well, stay consistent in the placement of your gear, and it will benefit you. That benefit may come as saved time, successfully achieved goals, or being able to aid others. Gear placement and consistency are essential but so is knowing your gear. Practice with your equipment, and learn how to operate it. Read the manual. On more than one occasion, I have been out in the field with someone who pulled out a shiny new piece of equipment they were using for the first time, only to find out that it requires an internet connection for the initial setup or registration before use.

Another similar scenario I have encountered is an investigator in the field with a piece of equipment that somehow gets the settings changed or tweaked. However, the operator has no idea how to get the equipment back to the needed setting. So, again, it's essential to know your gear and how to operate it.

During my law enforcement career, I was around to see many changes. One of those was the change from old school paper tickets to new, computer-generated, and printed citations. One particular co-worker of mine, who wasn't that much older than me, despised technology. He was not a computer person. I lost count of the times he would come into my office over the years with his laptop asking for help, many times for extremely simple tasks.

Don't get me wrong, I didn't really mind helping him; however, I found his attitude toward technology and his unwillingness to learn to be a bad example for some of the newer officers, but he was an old-school cop. He missed the old days of handwritten tickets and hand-drawn accident diagrams. However, I think I finally struck a chord with him one day.

I asked him, “Sarge, you know how to use your gun, right?”

“Yes.” He replied.

I asked, “You know how to use your radio, right?”

“Yes.” He replied, now visibly confused by my line of questioning.

“And you know how to use your taser too?” I asked.

“Yep.” He answered.

“Of course, you do; you’re a professional. But you know what? Your laptop and the programs on it are tools of your trade now too. They are part of the equipment and tools you need to use, like it or not. So, you should be working to be just as proficient with them as your gun, radio, and other equipment."

I don't remember his exact reply to my little speech, but I remember he walked off grumbling. However, I do think I got through to him as I noticed he made a concerted effort to solve his computer problems before asking for help, even bragging to me that he was able to solve some on his own.

My advice to you is to be professional. Learn about and know how to operate your equipment. I'm sure this advice seems redundant and a waste of time to some. But I assure you, much like the old sergeant, someone out there needs to hear it.

But what about types of equipment? I could literally write the rest of this book about the types of evidence collection equipment available, the pros and cons of the different kinds of equipment, and the vendors from which it could be purchased.

However, I will refrain from getting too deep in the woods on this subject (pun totally intended). New innovations are being made in evidence collection and equipment all the time. In fact, I'm sure that by the time this work sees print, there will be some new gadget, invention, or a previously unavailable resource to civilians that will be making a debut in the Bigfoot community.

Further, keep in mind that we all operate differently and have different ideas about approaching research. Some prefer to go into the woods with little more than a daypack and a camera, while others will lug heavy, cumbersome video and audio apparatuses miles into the woods. It all comes down to personal preference.

More sobering factors may come into play, such as fitness and health. How much stuff can you comfortably and safely carry? The most frustrating aspect for most of us is budget. What can you afford?

In the upcoming chapters, I will discuss the different types of evidence and some of the different ways to document and preserve them, including some pieces of equipment that could come in handy both in the field and in your home office.

However, I do have one more bit of gear advice. Do not sacrifice safety gear for research gear. If you are loading your pack and find yourself short on space or trying to shed items to save on weight, never leave safety gear behind in favor of research gear.

CHAPTER 4
PHYSICAL INVESTIGATIONS

For many, the most enjoyable aspect of Bigfoot investigation is the physical aspect. Hiking through the woods, alone or with friends, communing with nature, regrouping and recharging while spending time away from cell service, internet, and other interruptions. Many have entered the Bigfoot community on a quest to find Bigfoot and have remained as converted hardcore hikers and campers.

It is essential to spend time in the woods. To become familiar with the ordinary, everyday sounds of nature. Like the rookie police officer, the more familiar you become with your "beat," the more likely you will be able to tell when something is amiss.

I can explain the different types of evidence to you and how to collect, preserve and document that evidence, but I cannot teach you attention to detail. That is something you will have to develop on your own. One of my favorite stories regarding attention to detail occurred one night shift when I was training a new officer.

We were driving around on patrol, and the new officer was learning his way around and becoming familiar with landmarks and street names. We were driving down a particularly lonely backroad when I noticed the stop sign at the corner had been knocked down, apparently struck by a hit and run driver. I told my rookie we should get out and see if we could find any clues about what had happened and possibly fix the sign.

I got out of the patrol car, turned on my flashlight, and slowly began to saunter up to the sign. My new, impetuous rookie ran up to the stop sign, almost at a sprint, promptly stepping all over the tire tracks plainly visible in the soft dirt in front of the street sign.

Now, don't get me wrong, this was not the crime of the century, and I had little to no hope of finding the responsible party; however, my rookie had just marred up the evidence with his new patrol boots, essentially rendering any chances we had of finding the responsible party to almost zero. It would have been nice to have some documentation of the tire tread. In addition, if we found the offending vehicle down the road in a ditch, with an impaired driver passed out behind the wheel, it would have made putting the car at the scene of our property damage much more straightforward.

I used the incident as a teachable moment and told the rookie not to be so quick to rush upon a scene. Take your time, scan for evidence, and don't rush into a cold scene if you don't need to. Take your time and take it in. Hopefully, he learned something at that moment, and you can, too.

I have remarked several times, and probably will several more, on how similar Bigfoot investigation can be to police work. On the reality cop shows, they edit out the hours and hours of nothing happening. Only focusing on the exciting and entertaining parts. One thirty-minute episode of a reality cop show may contain the snippets of one or more weeks of activity. Same with Bigfoot shows.

I have often headed out into the woods with expectations of hearing wood knocks, whoops, screams, or even a visual sighting, only to return after an uneventful trip. It's been said many times before; "Best case scenario, you find Bigfoot, worst case, you have fun camping."

That being said, there will be times when things occur. For example, sounds of unidentifiable origin will be heard, impressions or footprints will be found on the ground, and even possible visual sightings and other events can occur when you are out in the field on a physical investigation.

One of the first steps toward correctly documenting and organizing these events is understanding the different types of evidence. Evidence comes in many different categories, and understanding the different classifications of evidence is one of the first steps toward correctly documenting, collecting, organizing, storing, and analyzing it.

CHAPTER 5
TYPES OF EVIDENCE

Physical Evidence: Physical evidence is perhaps the most all-encompassing category of evidence. It primarily includes anything tangible that can be picked up and handled. For example, in police work, it may consist of stolen items that have been recovered, a weapon used in an assault or murder, and tools used in a burglary to gain entry to a business or home. Pretty much anything you can touch.

Physical evidence can also be found during Bigfoot investigations, although not as often as in police work. Examples of physical evidence during a Bigfoot investigation can include sticks or branches that have been touched, moved, or broken by suspected Bigfoot and items in and around a camp that may have been touched, moved, or disturbed by suspected Bigfoot.

Some items can be more than one type of evidence. For example, a suspicious hair left behind at a campsite could be hair left behind by an inquisitive Bigfoot.

While the hair is undoubtedly physical evidence, it can be touched, picked up, and packaged; it is also considered trace evidence. Not only can it be considered physical and trace evidence, but it can also be analyzed and provide DNA evidence! Confused yet? We will get more into trace and DNA evidence later.

Transient Evidence: Transient evidence is perhaps, in my opinion, one of the most overlooked categories of evidence. It is possibly because it is one of the most challenging categories of evidence to document and collect. In essence, transient evidence is something that will go away.

Examples of transient evidence in police work can include ice cubes in a glass at a crime scene (they will eventually melt). In addition, smells like burnt gunpowder, perfume, and even freshly made coffee can help determine a particular scene's timeline.

When looking for Bigfoot, examples of transient evidence can include smells, footprints on soft, spongey, or wet surfaces, and fingers or handprints on windows or other items. Much of the evidence in Bigfoot investigation can be categorized as transient evidence because it is often found outside in the forest. Without documentation or collection, it will eventually be changed, altered, or destroyed by the elements.

<u>Digital Evidence:</u> Digital evidence is relatively self-explanatory. It can consist of an audio, video, or photograph file stored as digital media. Examples of digital evidence in police work can be footage from a security camera or even a photograph of the scene or evidence at the scene taken by an officer. Audio files can also be considered evidence. For example, all interviews were recorded on digital audio recorders and saved in the case files in my previous department.

Digital evidence in Bigfoot investigations can consist of many of the same types of files. For example, vocalizations caught on a digital recorder, video files captured on a trail camera, or photographs of prints, tracks, or a sighting location can all be considered digital evidence.

DNA Evidence: DNA Evidence is perhaps one of the most reliable and consistent types of evidence law enforcement agencies look for in investigating physical crimes. Television shows that focus on forensic investigations have painted DNA in an almost infallible light in the eyes of the public, and many jurors have scoffed at cases not involving some type of physical DNA evidence. DNA is simply cells that hold genetic information. That information is unique to each individual and family group.

In essence, DNA evidence consists of collecting a DNA carrier at the scene. Things like skin cells, blood, hair, and other bodily substances can be analyzed for DNA, and that DNA can be matched to a suspect.

In Bigfoot investigation, several instances of suspected DNA have been collected and analyzed with inconclusive results. In addition, suspected Bigfoot hair, blood, and saliva have all been collected and sent for analysis. I will touch on this more in the chapter on DNA but suffice to say, DNA, while a boon to law enforcement investigations, has almost been a bane to Bigfoot investigation.

Trace Evidence: Trace evidence can often include minuscule items that have found their way into a scene or onto a piece of evidence through contact or picked up from the environment. Usually very small, trace evidence in police work can consist of hairs or fibers from a suspect or suspect's clothing found on a victim. Pollen or plant fibers found on a victim's clothing can indicate if the victim's body was moved from one place to another.

Trace evidence in Bigfoot research can include suspected hair strands, debris that has been transplanted or disturbed on the ground, fibers, pollen, wood, or anything that could establish that a Bigfoot or suspected Bigfoot was in the area. Often trace evidence can be collected and analyzed for things such as DNA.

Witness Statements: Sometimes referred to as testimonial evidence, witness statements can arguably be said to be the bread and butter of Bigfoot research.

In police work, witness statements generally consist of the verbal or written record conveyed by someone who witnessed a crime, accident, or other events.

In Bigfoot research, witness statements can be retelling a sighting or other encounter story. There are literally thousands of different witness statements (of various levels of repute) available online, on podcasts, videos, and books available for consumption. Later in the chapter on interviewing and obtaining statements, we will examine the intricacies of verbal statements versus written statements and some of the science and biology behind what to look for when obtaining statements.

CHAPTER 6
EVIDENCE KIT

The first thing you will want to do if you plan on collecting and preserving physical evidence is to be prepared! Have the supplies you need on hand. Most evidence collection supplies can be purchased relatively cheaply. An internet search for police forensic supplies will lead you to any number of vendors on the internet who would be happy to assist you with purchasing some essential evidence collection equipment. Contrary to popular belief, most businesses will take anyone's money. You don't have to work for a police department to purchase paper or plastic evidence bags and other forensic supplies. You can even get custom ones with your Bigfoot research organization's name on them. Medical supply outlets can also assist when putting together an evidence collection kit. There is almost an infinite number of combinations of gear and supplies you can put in your kit.

If possible, I suggest having two kits. A big one that stays in your vehicle and a smaller one that goes in your pack or pocket. At a minimum, a good evidence collection kit generally contains;

1. PPE or personal protective equipment. This consists of gloves, a mask, and eye protection. These items will help protect you when dealing with possible biohazard materials such as blood, saliva, etc., and (if used correctly) prevent you from contaminating the evidence. You can also add more such as shoe covers, Tyvek suits, and hairnets.

2. Evidence packaging. This can consist of paper or plastic bags, boxes, plastic or glass jars, and even paper bindles. On television, most evidence is placed in clear plastic bags. This is so that the audience can see the evidence in the bag. However, in real life, paper bags are generally better, especially for items that may be wet or have blood or some other organic material on them. In addition, plastic bags can create condensation and damage or alter evidence, whereas paper bags breathe better and

allow items to dry and not get wet with condensation.

3. Swabs for collecting potential blood, saliva, and other miscellaneous evidence. Dried samples can be swabbed by adding a drop or two of distilled water to the tip of your swab. Be sure to package each swab individually after it's used and document where the sample was taken from.

4. Measuring tools. I also recommend a standard metal measuring tape and a flexible cloth measuring tape. Also, a small, flexible six-inch or so measuring stick can be kept in your notebook to quickly pull out for scale when taking photos.

From my time as a law enforcement officer, I can tell you that no matter how prepared you think you are, you will encounter a situation that you could not have prepared for. You will come across some evidence that will seem impossible to collect. I have had to collect items as tiny as hairs on the ground and as large as an entire automobile and recreational vehicle.

My best advice regarding evidence collection is to think outside the box. For example, one of my favorite out-of-the-box stories concerns a hit-and-run accident. An old sergeant I used to work with had been dispatched to a hit-and-run accident. Someone had wrecked a vehicle and fled the scene on foot, leaving the wrecked vehicle. The sergeant eventually tracked down the vehicle owner, only to have her claim the vehicle had been stolen, and the car thief must have wrecked the vehicle.

Now, sarge did not believe the owner and thought she had been driving when the vehicle was wrecked. He went to look at the car in the impound lot and found that there were a couple of spots of blood on the deployed airbag. Sarge thought about how to collect the blood from the airbag for a while. He couldn't swab it very well because the blood was literally soaked into the material of the bag. He couldn't just seize the airbag, as it was attached to the vehicle's steering wheel and not easily removed.

Thinking outside the box, sarge took out his knife and cut out the swatches of airbag stained with blood. Seizing those, he sent them to the crime lab, which determined that the owner's blood was on the airbag, placing her behind the wheel at the time of the accident.

That is just an example of how a little outside-the-box thinking can save the day when presented with challenges to collect evidence. While working in Alaska, we often found ourselves in a race to preserve evidence, such as footprints in the snow, as they quickly filled up with snow or began to melt away. Unfortunately, time and weather were rarely on our side. The same can easily be said about the search for Bigfoot.

While it's impossible to know exactly what type of evidence you will encounter out in the field, several types of evidence collection equipment are found in most kits. Items such as swabs, bags, PPE, and measuring tapes are standard issue items for Bigfooters. As previously stated, there are countless combinations of items you can assemble in your kit. However, I will suggest some more items that are not as commonly found in your standard Bigfoot collection kit and can be considered a bit "outside the box."

Some less than common items to consider:

- Hinge lifts/fingerprint powder and brush
- Blacklight
- Marker flags or tent numbers
- Collapsible hunting/bone saw
- Integrity tape for sealing bags
- Luminol powder
- Distilled water
- Lint roller/collector
- Sterile specimen cups
- Ink pen and permanent marker
- Magnifying glass
- Tags/labels with the chain of custody
- Work lights, both on stands and hangable
- Zip ties of varying sizes
- Garbage bags of varying sizes

Time and experience will give you ideas for even more items to include in your kit(s) as you progress in your search. Most Bigfooters worth their salt will have multiple kits for multiple occasions. A big one that stays in the vehicle, a smaller one that goes in a backpack for hikes and backcountry camping, and possibly an even smaller one that goes in a pocket for short hikes away from base camp.

CHAPTER 7
CASE NUMBERS

Now that you have a kit assembled, it's time to get out in the field and find some evidence. In this guide, I will convey some best practices and tips for collecting and documenting evidence. Often, when in the field, we rarely run into perfect conditions. Factors such as the weather, terrain, time constraints, and more can affect your ability to collect evidence. Our "thinking outside the box" will come into play during these times. You may have to get creative to collect, preserve and document evidence.

When collecting evidence for the police department, I would need a case number. Case numbers are numbers that are assigned to incidents and investigations so that they can be categorized, cataloged, and easily referenced. While looking into an investigation or sighting, I suggest you come up with some way of labeling your investigation to distinguish it from others.

Most police departments use case numbers like 19-0037. Generally, the number breakdown has the first number(s) in the case number referencing the current year. For example, by looking at the number mentioned above, I can deduce that this case or incident was reported in 2019. I always use the reporting year for the case number, not the actual year of occurrence. If you get a report year(s) after the fact, my advice would be to give it the number of the current year and then just document the actual date of the sighting in your investigation report. Then the second set of numbers will reference the incident number, so this will be the 37th incident or case file created in 2019. Using that number, I can find all associated files, evidence, and paperwork.

As you start your investigations, I suggest you develop some kind of system for organizing the information and evidence obtained during your adventures. You do not necessarily have to use the case number example I gave above, but some sort of naming and the documenting system will eventually be helpful.

It may be easier for you to use a location and date as your case number, for example, Bluff Creek 10-20-1967; this gives you a good reference point for where and when your investigation took place. I should mention that whatever system you decide to use, you should err on the side of brevity. Whatever your "case number" is, you will be writing it a lot depending on how much evidence you collect, so the easier to jot down, the better.

Further, you may want to use a case numbering system that does not give away too much information. Using location names, witness names, etc., may not be ideal if you want to keep those elements of the case private. Whatever system you decide on is up to you, but once you find one that works for you, stick with it. Consistency is essential, especially after a few years. Eventually, time has its way with all of us, and you will be amazed at what you will forget as you age.

Many times, I have found myself sitting around with my fellow officers reminiscing about "the good old days" when one of my co-workers would bring up a particularly action-packed, gruesome or humorous scene or call, and I would be astonished to find that I had forgotten all about it and had not thought about it in years! You know you are an old cop when you find yourself uttering phrases like, "Oh yeah, that's the time that one guy threw a toaster at my head. I'd forgotten about that." So, trust me, documentation is essential.

CHAPTER 8
EVIDENCE ITEM NUMBERS

After you determine how to categorize your cases or investigations, you need to decide how you will organize your evidence. You go out into the field to investigate a report and find a nice trackway of large, humanlike prints. You leap into action and get as many of the tracks cast as you can with your evidence kit. While casting the tracks, you come across a clump of reddish-brown hair hanging from a branch over one of the tracks.

From this one investigation, you now have multiple pieces of evidence. Much like using case numbers to distinguish cases and investigations from each other, it helps to develop a system to organize and categorize collected evidence.

You already have your case number for this investigation; it's 22-0015. So, all the evidence collected during this investigation will be assigned to case number 22-0015. But since you have multiple pieces of evidence, you need to be able to determine which part is which. There are a couple different ways to do this, and I will leave it up to you to choose which method you want to use but let me give you some examples that may help you decide.

The first is just to give the piece of evidence an item number. For example, the tracks will be numbered 22-0015 Item #1, Item #2, and Item #3. This method gives each piece of evidence its own item number under the case number. This method is suitable for investigations that yield lots of individual pieces of evidence. Theoretically, you could also use letters like 22-0015 Item A, Item B, etc. However, I caution against using letters because you are limited to the number of letters in the alphabet. Numbers, however, are infinite.

Alternatively, with items like tracks and track casts, you can employ a hybrid of both methods when you have multiple items that are the same thing. For example, the first cast in the trackway can be 22-0015 Item #1A, the second track can be labeled 22-0015 Item #1B, etc.

Then the clump of hair you collect would be 22-0015 Item #2.

Technically, all the casts are item #1, but you can differentiate between the different tracks by assigning a letter to the item number. After that, it is up to you to decide what method works for you and your team. Of course, you may decide on a different approach altogether. My advice is that whichever method you decide on, be consistent. Consistency will help you develop the mental muscle memory to perform the same way every time and lessen the chance for mistakes.

CHAPTER 9
PHOTOGRAPHING OF PHYSICAL EVIDENCE

Now that you have an evidence collection kit and a system for documenting cases, incidents, and evidence, let's get into the actual collection of physical evidence.

Once you find something that you and your team believe is worth collecting as evidence, the first thing to do is document it. Before you even think about touching it, you should take pictures with your camera. We are fortunate to live in the time we do, with the availability of digital cameras and storage; I once had an instructor tell our homicide investigation class, "Digital film is cheap. Take pictures until you are sick of it, then take some more." With the advent of digital cameras and digital storage, we can literally take thousands of photographs and store them on a device that physically takes up the space of a thimble. Truly incredible if you think about it.

The reality is that you only get one first chance regarding evidence. Before any item is collected, it should be photographed from every angle possible. Stand near it and photograph or video the site in a three-hundred-and-sixty-degree circle. I have seen many pictures from the scene of a death investigation or accident scene. The photographing officer took hundreds of photos of the dead body or cars involved in the accident but almost none of the surrounding area. It can be easy to get tunnel vision.

You may even want to take photos or videos from several yards away. Physically walk up to the object as your record/photograph. Even if the item you wish to collect is just a single hair, I suggest you document as much of the surrounding area as possible. Once you disturb an area and the evidence contained therein, you cannot go back in time. Therefore, it's best to take a few moments to document things as thoroughly as possible before moving an item.

If an item is disturbed before collection, it actually happens a lot, all is not lost; simply make sure you document how it was disturbed. Did a team member step on it? Accidently pick it up? It happens; no one is perfect. In your notes and report, make sure you document who disturbed the item, how it was disturbed, and the time it was disturbed. After photographing the object in its current location, have the person who disturbed it put it back where it was found, or as close as they can get it, and then photograph it in the place it was found. Again, just be sure all steps are documented through photos, video and notes.

After you have thoroughly photographed the item au natural, you or one of your team members should set some scale next to the item if possible. You should have some measuring tape or a small ruler in your kit for use as a scale. If not, you can use other items for scale, but, if possible, try to use something that anyone can replicate when using an alternative scale. My favorite alternative to rulers is paper money; all United States paper currency share the same dimensions, so anyone can copy your scale regardless if you only have a one-dollar bill or a hundred-dollar bill.

After the scale is in place, take more photos of the items from multiple angles. If possible, take photos with a scale showing length and width. Experiment with different modes of photography. Get to know your camera and how to use the macro focus for photographing smaller objects clearly. Further, experiment with photographing with and without a flash. In lower light situations, even when you can clearly see the evidence in your photographs without a flash, you may try taking some with a flash. Sometimes the extra light can expose things that aren't visible in non-flash photos.

One of the most valuable techniques I've found when it comes to the photographing of evidence is oblique lighting. While relatively simple, oblique lighting can reveal details otherwise hidden from the naked eye.

Oblique lighting uses a light source at a low angle shining on a prospective piece of evidence. So, if you find a print on the ground, you would take a flashlight or other light source and shine it across the track; personally, I like starting at a ninety-degree angle and photographing the track. The light can show details and characteristics of the track that are otherwise invisible to the naked eye by casting shadows.

I have found the technique of oblique lighting to be beneficial in discovering footprints and fingerprints found on flat surfaces. For example, while investigating more than one burglary, I have found that shining my flashlight on a table or counter has revealed a shoe print or handprint that I could either photograph or use a gel lift on.

Get creative when it comes to using your flashlight. Don't just think that you see all the details you can by shining your light on an object head-on. Instead, shine your light on things at angles and move the light around. You will be surprised at some of the details that oblique lighting will reveal!

Most modern smartphones have excellent cameras capable of taking high-definition photos and videos. There are many tutorials and videos online that can teach you about the features and capabilities of your phone cameras. I suggest becoming somewhat familiar with them before heading out into the field. Practice around your house and in your yard. Try to get good high-quality photos of birds in the trees and dog tracks in your yard. Get comfortable with your equipment.

I worked with an officer who was good about taking photographs at crime scenes, but his photos were almost always blurry. I could literally tell his photos from the other officer's because most of his photos were blurry and out of focus. At the time, the digital cameras we were issued required you to hold down the shutter button for a second so the camera could auto-focus. Then, when the lens was focused, you just pushed the button down to take your picture. I pulled the officer aside one day and told him about the autofocus feature and that if he just pushed the button down a little bit and waited for the camera to focus, he would take much better pictures. His response was something along the lines of, "I know. When I push the button, the camera should just take a picture. It's not my fault the cameras suck." Blaming his poor photography skills on the quality of the equipment issued by the department.

Not long after, the department stopped issuing cameras and switched to smartphones. The officer's blurry photos improved. Unfortunately, the smartphones were smarter than some of the less-than-tech-savvy officers, which unleashed a new batch of problems. There are two things that police officers hate more than anything else, change and the way things are.

CHAPTER 10
PACKAGING PHYSICAL EVIDENCE

After you feel you have thoroughly documented the scene with photographs and are ready to collect your evidence, I suggest you keep a log, like a crime scene log. Document in your log who collected what piece of evidence, who handled the item, when the item was collected and where it was collected.

When it comes to determining where a piece of physical evidence is located, you can use several different methods. In addition, the type of environment you are operating in may influence which method you use to establish evidence item locations.

For example; police collecting evidence at a crime scene may pick up a bullet shell casing and simply put the location as "kitchen floor," or they may break the room down into grids and say the case was found in grid A 12 or "kitchen floor, northwest corner, next to fridge."

Obviously, indoor scenes can be easier to manage, location-wise, than outdoor scenes. However, when dealing with outdoor scenes, especially scenes in an area with no easily discernable landmarks, like a fire pit or campsite table, the only way to mark a location may be with latitude and longitude. There are several apps available for smartphones and tablets that allow you to take photos with date and time stamps as well as latitude and longitude information displayed on the photos.

Sometimes there can be an option other than lat/long for documenting location when processing an outdoor scene. For example, officers will sometimes find a “Reference Point” or RP for short when investigating traffic accidents. Because most roads and sidewalks are pretty non-descript and plain, officers will see something nearby that is not likely to be removed, changed, or taken away anytime soon. Examples of RP’s can include but are certainly not limited to fire hydrants, lamp posts, traffic control devices, signs and manholes. By documenting the involved vehicles’ relationship to the RP, officers then have the means to "re-create" the scene and have the measurements for scale.

In an outdoor setting where you find something of evidentiary value that needs to be collected and documented, you can attempt to locate a reference point if the terrain permits. Perhaps a unique-looking tree or rock, preferably something that will not be easily moved and will be there for a long time. You can measure the distance between the reference point and your evidence to be collected to establish the exact location at the scene. Further, I suggest photographing the evidence with the reference point in the photo. You should mark the reference tree with a trail marker or orange tape for the photos.

Ideally, when you finish documenting a scene, a complete stranger with no knowledge of the area should be able to come in and recreate the scene using your notes, measurements, and photos. However, suppose you want to give the scene the complete documentation treatment. In that case, you may wish for one of the more artistic team members to sketch a diagram with measurements between items, although some may argue that an excellent, thorough video, may supplant the need for this. If possible, this may also be accomplished with overhead shots from a drone. This level of detailed documentation may not necessarily be needed for each scene, but I would be remiss if I did not include it here. Remember, take what you need, put it in your toolbox and use it when necessary.

After you are satisfied that you and your team have adequately photographed and documented the evidence's location, it may be time to physically collect the evidence. Donning the appropriate personal protective gear and preparing whatever receptacle for your evidence should be performed in an area to ensure that you will not disturb the evidence.

Do not shake or fan open your paper bag while standing over your potential evidence. You do not want to inadvertently disturb any potential hair or other trace evidence, and adversely, you do not want to distribute other dirt and debris onto your evidence before you collect it. Further, do not don protective equipment standing over or close to potential evidence. Again, you do not want to inadvertently contaminate it with any sweat, hair, or other items that may fall from your person.

Depending on the type of evidence you are collecting, you may be able to physically handle it in several different ways. In the case of a larger piece of evidence, you may be able to just pick it up with your (gloved) hand. On the other hand, you may be able to collect the item with little or no physical contact with the item at all. In some cases, you may be able to just put your evidence receptacle over the item and pick it up by touching the bag and not the item itself. This method can be beneficial if the item has the potential to yield DNA evidence. If it is smaller evidence such as hair or other trace evidence, you may need a small tool to help you collect it, like tweezers.

If you collect an item with tweezers, I suggest, if possible, sterilizing the tweezers between uses. This will help prevent cross-contamination between pieces of evidence. A dab of rubbing alcohol and a cotton ball can usually do the trick; glass cleaning wipes are a handy substitute. Just make sure the ones you use contain alcohol, be sure and give the tweezers some time to dry between uses. Further, if you collect multiple but similar items from different areas, do not package them all in the same container. For example, do not place ALL the hair found near an area in one bag. If you find hair near a campfire and more hair near a tent or several feet away, package the hair(s) separately and document where each was found.

If possible, I like to label my evidence receptacle before placing the item in it. This makes it easier to write on and less likely that my pen or marker will pierce the bag and contaminate the evidence. I have never had any issues using a black permanent marker to label my evidence bags. However, if you do not like using markers on your bags, you can certainly use pens or even purchase custom stickers with fill-in-the-blank fields to place on your evidence bags. I discourage using pencils; you want the information to be as prominent and permanent as possible. If you order bags from an evidence supply company, many will have fill-in-the-blank fields already on the bags.

When it comes to labeling your evidence bags, you want to convey as much info as possible about the item inside the bag without writing an entire report on the bag. I suggest the following format:

Case Number: 22-00015
Item Number: Item#5
Date: 10/22/2022 (1450hours) *date and time collected*
Name of the person who collected the item.
The location where the item was collected.
Description of item. Example: 6-inch-long reddish-brown hair collected from a tree branch above a suspected trackway.

You can decide how many of the above fields should be used on your evidence bags. Once you have your bag labeled and the item safely in the bag, you should seal the bag so that the item is not lost from the bag during transport. Personally, I like yellow or red tamper-evident tape. This is a special tape that can be purchased, and when it is applied to a surface and torn, it's nearly impossible to re-seal so that the broken seal cannot be hidden. If the seal is broken, it is pretty evident.

An extra step that I liked to take, but is totally optional to you, is when I would seal my evidence bags with the tamper-evident tape, I would sign the tape/bag with my initials and the date. I would do this so that my writing was on both the tape and bag. Then, if the tape was tampered with, it would be evident because my signature and the date would no longer be lined up.

The type of bag you use may determine how you decide to seal them. For example, most plastic evidence bags have adhesives on the flap that you can expose and seal the bag with. Paper bags can also have the lip folded and taped or even stapled shut. Before sealing any bag, keep in mind that you are not damaging, contaminating, or otherwise compromising the evidence contained within the bag.

Paper bags are typically more suited for evidence collection than plastic, as I mentioned earlier. However, plastic bags can and do indeed have a place in evidence collection. Despite being prone to condensation, which can damage evidence, plastic bags are often lighter weight and take up less space than paper bags, making them an excellent candidate for a small, portable collection kit.

Any item you collect that is wet or moist, if possible, should be placed in a paper evidence bag. When the items are transported back to your base of operations, the wet/damp items should be taken out of the bag and allowed to dry. This includes any type of moisture. For example, if the items were wet from the rain, recovered from a stream or lake, or even if the liquid is blood, you should give the items time to dry before putting them into long-term storage or sending it out for potential analysis. Leaving a moist/wet item in a bag for too long can lead to the growth of mold or fungus and risk the moisture weakening the integrity of the container the evidence is stored in.

Suppose you have to remove an item of evidence from the packaging, perhaps for drying, or allow someone to inspect it or a potential specialist to analyze it. In that case, I suggest that whatever surface you are using to unpackage the item, you cover it with white or brown butcher paper. This will not only protect whatever surface you are using from being in contact with the evidence, but if any trace evidence, like dirt or hair comes off the item, you will be able to see it on the butcher paper and collect it if needed.

If you are in a situation where you find yourself bringing out multiple pieces of evidence for viewing/analysis, I suggest bringing them out one at a time and switching out the butcher paper underneath to prevent cross-contamination.

Remember that when it comes to collecting physical evidence, think outside the box. You may have to resort to unconventional methods to get your evidence safely out of the woods or away from the scene. From wrapping half a car bumper in large garbage bags to chain sawing a limb off a tree, you may have to literally invent ways of obtaining what you need to further your investigation.

CHAPTER 11
STORAGE OF PHYSICAL EVIDENCE

One of the challenges of collecting physical evidence is where do you put it? Unfortunately, most of us do not have access to a large, environmentally stable, humidity-controlled storage space. As a result, many citizen scientists have probably stored valuable, significant scientific discoveries in places like desk drawers, hall closets, and even refrigerators, right next to their beer and pickles.

You cannot go wrong with a cool, dry place like a closet or garage for most items. However, when storing evidence items, you should consider a few factors. Namely, will the item be safe? Also, is the item in danger of being contaminated?

I'm not just talking about cross-contamination of the items when I mention contamination. I'm also talking about the item being stored improperly; if the items contain some organic matter and it is left for too long under the wrong conditions, the organic components can begin to decompose and break down.

Safe is also a relative term. Are the items out of the reach of curious children? Spouses, who are spring cleaning? If you are going to store evidentiary items in your residence, I suggest that you make sure all the other residents are briefed on the items and told not to disturb them. If possible, keep them in a secure location with limited access to a minimum number of persons.

The best way to store evidence is to not store it at all and to get it to where it needs to go. First, find the best resource available to analyze the evidence you have collected, be it physical evidence or DNA evidence. Then, get it out to be analyzed as soon as possible. The longer you hold onto a piece of evidence, the more chance it will get lost, stolen, contaminated, or even forgotten. Getting evidence in the right, qualified hands and getting it analyzed and studied in a safe, timely manner, is the most efficient course of action.

Often, the best person for the job of evidence analysis is not going to be within driving distance of you. Depending on the type of evidence you have collected, there are acceptable ways to transport your item to be analyzed.

We were about a four-and-a-half-hour drive one way to the State Crime Lab at the department I worked at. Obviously, we could not spare the manpower or the overtime to pay someone to drive eight hours to drop off evidence at the crime lab. So, we would simply use the United States Postal Service for most items.

The State Crime Lab had standards for the packaging and shipping of evidence. These standards ensured safe and secure transport through the mail to their destination.

First and foremost, you do not want your evidence to inadvertently harm some innocent postal employee; secondly, you want your evidence to arrive safely at its intended destination. I will lay out the most common way my department would send evidence to the crime lab to be analyzed. This is by no means the end all be all final word on the subject. There may be other entities with better ways to transport valuable evidence. These are just my best recommendations based on my experiences.

Now, these are guidelines and obviously will not be able to be applied to every single piece of physical evidence you collect but will work for the majority of items.

Get your item out of its secure storage location. Leave the item in the evidence collection bag/box or whatever container you have stored it in. You do not want to risk contamination by handling the object. Obviously, there are exceptions to almost every rule. However, suppose the item(s) is something potentially dangerous, like a knife or other sharp object. In that case, you may want to take steps to render it safe for transport, like putting a cardboard sleeve around a knife or sharp object so it does not pierce the box it is being transported in.

Find a box of appropriate size. Ensure the box is relatively clean and do not use containers used to transport biological material. For example, do not send out an evidence item to be tested for DNA in a box that you received mealworms in for your pet lizard.

Also, suppose your item(s) are of a biological nature. In that case, you should find biohazard stickers to affix to the box so that anyone opening the box knows to don the appropriate PPE.

Securely pack your evidence in the box. Make sure the item(s) will not be jostled around too much, and use packing material if you need it to help secure the items. Once you are confident you have the items securely packed, it is time to seal the box. Before sealing the box, make sure you have included any reports, paperwork, or other documentation that your destination may have asked for. Some labs may send you a sheet with a bar code on it to include in your package so they can scan the item into their inventory when they receive it.

Be sure to document in your notes, field report, or both when the item was shipped, how it was sent, the tracking number, and the item's final destination. Likewise, when/if you get the item back, document that information in your notes or field report.

Once you are sure the package is ready to be sealed up, tape it shut with some good, reliable packing tape. Then, wrap the box up in brown butcher or packaging paper using the packing tape. Now, once the package is wrapped in brown parcel paper, what you want to do is get a good strong water-activated tape. The type I always used was brown like the parcel paper and just needed a damp washcloth or towel run over it to activate the adhesive.

Take the water-activated tape and tape over all the seams on your parcel paper. Make it so that they will have to cut or tear the tape off if anyone wants to get into the package. Do not get the tape too wet because you do not want to weaken your parcel paper; you just want enough to activate the adhesive and stick to the box. The water-activated tape creates a strong seal along the seams and creases in the parcel paper. If the tape is broken or tampered with it is easy to tell.

Now you are ready to take your package to the post office. Obviously, you do not have to use USPS to ship your evidence, but I believe they offer the best service for the job, certified mail. When you send your evidence, you should ship it by certified mail. While a bit expensive, this method gives you a tracking number and a return receipt when the item reaches the intended destination with a signed receipt. The signed receipt can be helpful because it gives you the person's name who actually received the item from the mail carrier.

Before mailing any evidentiary items out, I suggest that you obtain some sort of acknowledgment from the lab or individual that you intend to send it, that they are agreeable to receiving it and that they will be looking for it in the mail. This suggestion applies tenfold to items of a biological nature. Do not…and I cannot stress this enough, do not send surprise poop in the mail. Always make sure the evidence destination, be it a laboratory or private citizen who is a subject expert, knows exactly what you are sending them in the mail.

When I was a young officer, I was told a story about a town in northern Alaska where a burglary had occurred, and the burglar had defecated on the floor in the building he had burgled. Some dedicated, eager rookie officer had dutifully collected the feces and sent it promptly to the crime lab to be analyzed to find the identity of the incontinent lawbreaker.

Needless to say, the crime lab was not happy to receive their box of poop. This surprise mail drop culminated with a strongly worded email from the State Crime Lab to all law enforcement agencies, dictating that feces would be sent to the crime lab only in the direst of circumstances and only with the approval of the head of the lab.

I told you that story so that I can tell you this one. Many years later, I come in for my shift and get dispatched to a local office building that had been burgled during the night. I meet with employees at the door, and they walk me around the building showing me items that were disturbed, damaged or stolen.

The tour culminated in the office of the head of the branch where the burglar left the company a big brown surprise on one of the guest chairs in the nicest office in the building.

Like a good officer, I photographed the feces and the room, looked for other clues, and departed the office. My then sergeant at the time hung around the building after I left and was chatting with some of the employees he knew in the building. Later, after we had met back up at the police department, he informed me the staff of the office building were critiquing my investigative skills and were not happy that I had not collected the feces in the chair. I was told one employee had remarked, "So much for CSI." With an eye roll as they marched toward the boss's office with a roll of paper towels, rubber gloves, and disinfectant spray.

Now the sergeant and I both knew that collecting the feces would not help us in our investigation. Frankly, we were both a bit bemused that anyone would think it was a good idea, but we were both familiar with the "CSI effect" on the citizenry of our jurisdiction, and we shrugged it off. We will discuss this phenomenon in more detail later. For those of you wanting to know what happened with that case, I identified and arrested the burglar by the end of that day and recovered the stolen items. All without having to collect his poop. Imagine that.

CHAPTER 12
MANAGING DIGITAL EVIDENCE

The introduction of digital evidence is arguably one of the most dramatic changes to the discipline of evidence collection in the last few decades. Having had access to my department's evidence room, I was able to see how old case files were assembled and stored. Old homicide cases would be packed in boxes. Often, more than one box per case, with binders full of reports and photo albums full of pictures from the scene.

In modern times, most paper case files are purged after a few years. The paper reports and documents are scanned and saved with any photographs, videos, and any other evidence that can be saved digitally. Today, in the space that one homicide case would take up in two banker boxes filled with reams of reports and photos, you could literally fill that space with external hard drives capable of holding terabytes of information. You could fill the same physical area with machines that can hold millions of photos and reports without exaggeration. You can now store millions in the same physical space that once had one bloated physical case file. Truly incredible if you really think about it.

However, as with all new means and methods, there are pros and cons. There are always those suspicious of new ways of doing anything, and there is nothing inherently wrong with that. However, the pros of digital storage and the cons should be weighed. One physical copy of an old homicide file taking up half an evidence room shelf is not without risk. Pipes burst, buildings catch on fire, and natural disasters happen. That one homicide case being stored on a shelf can be destroyed and rendered useless just as easily, if not easier, than a digital copy of the same case. In fact, the digital copy of the same homicide case can be saved on a thumb drive or external hard drive and copied onto a backup server, creating more than one copy of the same bloated case file that takes up almost no physical room.

If you find yourself investigating the Bigfoot phenomenon, you are likely to wind up with some form of digital documentation. It may be a photo, audio recording, or even video.

Later on, when we get into case file management, I will go into more detail on this. Remember when I talked about case numbers? When you open a "case" or investigation, I suggest you get on your computer and create a file for your case. Usually named after your case number. Then, I like to create separate files for audio, photographs, and/or videos inside that file. Inside these subfolders, I like to store my digital evidence related to that particular case.

In the modern world we live in, technology is changing and evolving at a breakneck pace. As a result, some of the things I write about in this chapter may be obsolete soon, if not already. So, again, take what you can use out of this toolbox and leave the rest.

When I first started in police work, we did not have body cameras. Instead, we would all carry small digital audio recorders at the departments I worked for, usually in a little pocket on our vest carriers. Upon arriving at a call for service or contacting a citizen during a traffic stop, we would turn on recorders and record the audio of the interaction with the citizen. Sometimes, our contact would lead to an arrest or the issuance of criminal charges. If this happened, we would take our recorder back to the police department and upload our audio files from the contact onto the department's server to be held in the case file.

These audio files would eventually be burned onto a disc and mailed to the District Attorney's Office along with the paper copy of the report. I was told, when I started, not to change the name of the audio file when I uploaded it to the server. Our little audio recorders would give the file a name like WS000192. The officers that trained me told me not to change the name of this file; the reason was that if the case were to go to trial, and the defense attorney was asking you questions on the stand, he might ask something like;

"Officer Baxter, did you alter, in any way, the audio recording file you sent to the DA's Office?"

Now, technically, if you renamed your audio file from WS000192 to something like, "Interview with suspect", you altered the file. You changed something about the file, even though it was minor and was for the convenience of yourself, the DA, and anyone who had to look through or listen to the audio associated with the case. So, technically speaking, you have to answer "yes" to this question.

Defense Attorneys are notorious for asking questions like this and not allowing you to explain. I've often been asked a question that required more than a yes or no answer and been told by the Defense Attorney to answer the question yes or no. They do not want you to explain yourself because they only wish the jury to hear a partial answer that they can use to influence the jury's decision.

So, in my training officer's instruction, I was told not to tamper with the name of the audio files. Made perfect sense to me, although I did see how it could be made finding the correct audio you needed difficult. I eventually found a way around this, and when I would write my report, I would note that this particular contact was recorded and saved in the case file as audio file

WS000192. This made finding a specific bit of audio more convenient.

Fast forward a few years, and our administration informed us that we would save our audio as the case number and then a description of the audio from this day forward. An example would be 22-009 (Interview with a suspect). A few other officers and I spoke up and said that we disagreed with this directive because of the reasons I brought up earlier and how it would make us have to answer "yes" to the question about altering our audio files. We were basically told not to worry about it, and the DA wanted to do it this way.

For the rest of my time working in law enforcement, I expected this to come up in court someday, with me having to answer yes to the question about altering my audio files. Much to my surprise, it never did. I often wonder if the DA's Office and the Public Defender's Office came to some collective agreement that it was easier for everyone involved if they let the officers get more descriptive with their audio file names?

Effectively coming to a truce over the matter so that the Defense Attorney would not hassle us over it?

I bring all that up to show you an example of how digital evidence can be scrutinized and nitpicked. Something as innocuous as a cosmetic name change for convenience can be used to besmirch the veracity of digital evidence. Granted, your digital Bigfoot evidence is not likely to end up in a criminal court case. However, I can confidently say that any digital evidence you obtain, especially video, will be scrutinized just as much, if not more, than any digital evidence in a criminal court case.

I have a few suggestions for how to handle digital evidence. Of course, they will not protect your evidence from the horde of scrutiny it will receive should you decide to release it, but it will help you keep track of a digital chain of evidence.

It can be hard to convey the best way to transfer your evidence from its collection point; audio recorder, smartphone, video recorder, or thermal imager, to your case file. Some devices have removable storage like memory cards, while others have built-in memory, and files must be transferred via a cable or even Bluetooth.

Regardless of how your evidence gets moved from your device to your case file, you should note it in your notes or your report. Further, if your evidence is particularly impressive, I suggest, if you can, keeping the file intact on the memory card or on the device. Log your memory card in as evidence and save a copy in your case file if you feel that the original unaltered file may be needed in the future.

Most of us cannot afford to keep all our memory cards in evidence and replace them, so we upload our memory card content to our computers and case file. Once you transfer your evidence, be it audio or video, I suggest you keep an original version in the case file as evidence. Then, if you want to make a couple of copies, keep them in a separate file labeled, "Working Copies", these are the copies you can edit. Maybe they need to be shorter; perhaps they need to be enhanced?

Perhaps your original video file is over an hour-long, and you only have a few seconds of footage you want to show other people? Instead of finding a way to send over an hour's worth of footage to people, or posting the whole thing to social media, take your working copy and edit it down to the relevant footage. This maintains your original video and leaves you with a working copy that you can pass around.

I know what some of you are thinking right now. "But Beans, if I am working with copies of my video and audio, aren't they poorer quality than the original?"

The short answer is: No. Generally, digital copies of audio and video do not lose quality when copied. However, if you convert your video or audio to a different format, then changes in quality can happen. Remember in the previous chapter when I talked about learning how to use your gear? This is where that advice can really come in handy. Know what format your device records in. Some allow you to choose what format your data will be stored in. Research the options available to you, and whenever possible, always record in the highest quality available. Better quality generally means bigger files; however, a few pixels can make all the difference in the search for Bigfoot.

I could probably write multiple chapters on what to do with your digital evidence. There are any number of Bigfoot "experts" that you can send audio and video to, and they will give you their opinion. Nothing wrong with that at all; however, I urge you to be cautious and make sure you thoroughly vet the person before handing over any evidence of any kind. It would be a terrible scenario indeed to send someone evidence that you worked hard for only to have them claim the evidence as their own or to "accidently" release the evidence so that it appears to have originated with them. Unfortunately, these types of scenarios can and do play out.

Also, I would be remiss if I did not mention EXIF data regarding digital evidence. EXIF is short for exchangeable image file. While the existence of EXIF data is not exactly common knowledge, it is not necessarily a secret either. EXIF data is attached to most photos and videos taken with modern devices, especially smartphones. The data attached to the images and videos can include all kinds of information about the photo and the device that captured the photo. You can see the resolution the photo was captured in, the date and time, whether or not the flash fired when the photo was taken, the make, model, and sometimes the serial number of the device that took the photo, and most importantly, the latitude and longitude of where the photo or video was captured. Can you imagine the implications if we had that information on the Patterson-Gimlin film?

If you are given photos or videos from a witness, I recommend running them through an EXIF data viewer. You can use the information from the EXIF data to collaborate or disprove the witnesses' claims. Any internet search for EXIF viewer should bring up any number of reliable, safe sites in which you can upload your photos and view the data.

While EXIF data can be beneficial to an investigator, it can also be something to beware of. Remember when sharing your own videos and photos that anyone with knowledge of EXIF data can take advantage of it and find the location of where a particular image or video was captured. This can be particularly troubling if you are trying to keep a location secret to keep out trespassers or keep the area from being flooded with other Bigfoot enthusiasts.

However, do not let this information completely scare you off the idea of sharing your photos and videos on social media. Most major social media platforms scrub the EXIF data from photos and videos posted to their platforms. This helps maintain their users' privacy and is, in my opinion, an intelligent move. For example, you wouldn't want someone to be able to find the location of your house or workplace just by running a photo you posted through an EXIF viewer.

I have a dear friend in the Bigfoot community who I often run through different "what if" scenarios. Almost without fail, most of these scenario-based exercises devolve into, "What if we get the next Patterson-Gimlin film?", to which my dear friend's answer is always the same; "I tell you what I won't do with it, you won't see it on [expletive deleted] Facebook, that's for sure!"

I disagree with my friend's take on this answer, but not entirely. If I had some excellent video footage that I felt was the equivalent of the Patterson-Gimlin film, the first thing I would like to do would be to get it seen by as many respected, educated people as I could. Not necessarily people in the Bigfoot community, either. I would try to think outside the box and show it to experts and professionals in fields that you would not necessarily associate with Bigfoot and cryptozoology.

However, after I felt the film had been scrutinized by as many in academia and Bigfoot as needed. I would absolutely love to share it on social media. I say that for a couple of reasons. The first is, that it would be great to have something special to share with the world. The second is, I would be curious just how well it could hold up to the scrutiny of the internet. No doubt some would dismiss it outright as a hoax or a fake, but some people out there would put it through its paces and try to break it down and debunk it. I would love to capture something that held up to that type of scrutiny.

Further, if you are working in a group or a team, before even going out into the field, I urge you to sit down with your teammates and discuss what to do with evidence collected on team outings. It might even be a good idea to have an agreement or waiver on file signed by all team members as to the ownership and rights to any evidence collected. But, again, before you even step foot into the field, you should make sure everyone is on the same page about handling any potential groundbreaking evidence collected.

In conclusion, when it comes to digital evidence, always take steps to preserve your original copy of whatever file you have while creating "work" files that you can edit, distribute and otherwise alter. Think outside the box when it comes to video analysis and be cautious about whom you share your digital evidence with. When it comes to digital evidence storage, I like to adhere to the adage, "Two is one, one is none." Keep your files on at least two devices; if possible, keep one in a separate location from the other and in a fire-resistant container or safe.

CHAPTER 13
DNA EVIDENCE

Anyone who has consumed any offering from pop culture in the last few years has almost definitely heard of DNA. DNA has replaced radiation as the science of choice in science fiction in recent years. In years past, many super-heroes obtained superpowers from any number of accidental exposures to radiation, from being caught in a radioactive blast to being bitten by a radioactive bug. Radiation was the explanation for fantastic mutations, powers, miracles, and monsters.

However, more recently, the fictional radioactive accidents from years gone by have been replaced by accidents involving genetics and DNA. Instead of a radioactive bug passing on superpowers, the origins are now being told with a genetically altered bug that changes our hero's DNA, giving them the power of an insect. Power-inducing explosions have been replaced by DNA-altering formulas and events. Our fiction is evolving to include our findings in the real world.

While DNA was technically discovered before the first nuclear bomb was constructed, the actual application of DNA testing did not come along until much later. So, to put it in context, I am (and some of you are) actually older than DNA testing. This makes the science of DNA a relative newcomer in the scientific arena.

Without getting too deep into a scientific explanation, DNA stands for deoxyribonucleic acid. It is found in nearly all living organisms and is responsible for housing genetic information. I am certainly no expert on the intricacies of DNA. However, I was trained to collect it and send it to the lab. Anything more than that was handled by lab staff and expert witnesses. Any case in which DNA was a factor in the evidence usually resulted in an expert witness being sent from the crime lab to answer the technical questions in court.

As the study of DNA has progressed over the years, it has been discovered that it can be used almost like a fingerprint to identify people and even help identify health risks and problems that people may be prone to.

Every year, some new facet or use for DNA is discovered.

Not long after the aspect of genetic fingerprinting was identified, it was used to solve a series of horrific murders in Britain. There is an excellent book on those events by American author Joseph Wambaugh. If you are interested in forensic biology, I suggest checking it out.

As with all new emerging science and technologies, I'm sure some Bigfoot enthusiast threw their hands up in the air upon hearing about DNA and DNA sequencing and said, "This is it! The hunt for Bigfoot is over!", genuinely believing this new science would lead to the imminent and inevitable discovery of a new species.

Alas, this was not the case, and the same scenario has played out with the advent of many other technologies like thermal imaging, drones, and motion-activated trail cameras. Unfortunately, while all seem to be possible game-changers in the search for Bigfoot, and some have produced some compelling, interesting evidence, none have brought in anything that will officially cement the existence of our elusive, bipedal friend.

In my law enforcement career, I was fortunate enough to have some experience in collecting DNA evidence and the use of DNA evidence in solving cases. Using DNA to establish someone's presence or involvement in a crime scene is quite different from proving the existence of a new species. For starters, you know you are looking for human DNA in law enforcement. In most cases, you are looking for some sampling of DNA to collect at a scene so you will have a sample that you can either compare to a suspect or run through a DNA database and hope to identify a suspect.

My good friend, world explorer Adam Davies traveled to Sumatra in 2001 in search of the elusive orang-pendek. Similar to Bigfoot, the orang-pendek is unrecognized by science, a bipedal primate said to live in the jungles of Sumatra. Adam and his intrepid companions found tracks that their very adept tracker could not identify and strange hairs near the tracks.

They dutifully collected this evidence and brought it out of the jungle to be analyzed.

Hans Brunner, an expert on mammal hair, was tasked with analyzing the hairs Adam and his compatriots found. Brunner was a crucial figure in the acquittal of Lindy Chamberlain. Lindy was convicted of murdering her infant daughter, who went missing on a camping trip in the Australian outback. Lindy insisted on her innocence and claimed a dingo had taken her baby. After finding and identifying dingo hair on the baby's clothing, Brunner was instrumental in getting Lindy acquitted.

David Chivers, a primatologist at Cambridge University, analyzed the tracks cast by Adam's group. Both Brunner and Chivers declared that both pieces of evidence belonged to an unknown primate species.

The hairs were consumed during the analysis process and were no longer available for further testing. Samples of rattan bush that Adam suspected the orang-pendek was eating and hoped may contain trace amounts of saliva were sent to Dr. Todd Disotell, who could not extract DNA from the sample.

While frustrating, this is a scenario that I often ran into during my law enforcement career. For example, a cigarette butt or other seemingly DNA-rich item would be collected and sent to the lab, only to receive the lab results, "Insufficient amount of DNA available for analysis." As a result, a seemingly crucial piece of evidence would often be nothing, taking an investigation back to square one.

Television has not helped. Shows like CSI and countless true crime documentaries have implanted the idea that DNA should be a part of every investigation in many potential jurors' minds. Petty theft and shoplifting cases are being scrutinized for not having any lab work or DNA even when there is a video of the theft!

Hans Brunner's work with the hairs from the Chamberlain case was enough to overturn a murder conviction in a court of law but not enough for science to recognize a new primate species. As a result, people have been sent to prison for life with less evidence than the amount of evidence that supports an unknown primate's existence. Think about that. Is it the justice system or the scientific method that needs revamping? Or both?

As Adam's experiences tell us, just because you get the evidence doesn't necessarily mean you get the DNA. Many prominent figures in the Bigfoot community have asserted that DNA will not be enough and that someone will have to produce the source of the DNA, a body. Even more cynical enthusiasts have claimed that not just one body is needed for discovery but that two bodies will have to be produced and delivered to science to facilitate discovery. If one lone body of the target species is produced, the scientific community will dismiss it as a genetic anomaly or a freak. Ideally, a mating pair would be best to prove the species' existence. The male and female pair would go toward establishing a species out there capable of reproducing.

When it comes to DNA evidence, my thoughts often turn to the number of suspected Bigfoot DNA samples that have been sent off for testing only to come back as "contaminated by human DNA." I am no expert on DNA and its intricacies, but I have wondered more than once if it is not so close to human DNA that it gets mistaken for human DNA? Experts in the scientific field tell us this is not likely, but I cannot help but wonder. If the DNA we are looking for has not been officially recognized and cataloged, how can we confidently say it wouldn't be nearly impossible to distinguish from human DNA?

This may give the impression that I am not endorsing the collection of potential DNA evidence. This is not true. I believe that for discovery and scientific recognition to occur, we must use every tool at our disposal. Even tools that have not necessarily been successful in the past.

One of the hurdles with DNA evidence is finding somewhere that will analyze the evidence you collect. Some commercial DNA labs have services related to animal DNA identification available. Still, before spending any money or sending them any suspected

DNA, I would contact them and have a frank conversation. Determine if the services they provide are adequate for recognizing and sequencing possibly new undiscovered DNA and if those services are within your budget. Use your discretion to determine whether you should use the word "Bigfoot" in these conversations. In some circumstances, it may generate curiosity and interest; in others, it may lead to the termination of the conversation.

Thinking outside the box can also apply to DNA evidence. It cannot hurt to reach out to local veterinarians, zoos, and college campuses to see what kind of resources they can help you with. College campuses especially may be interested in looking at DNA at little or no cost. However, with that comes the stigma of a potentially groundbreaking discovery being overlooked because it was made by a biology student and not given the pomp and circumstance it deserves. So, again, have a conversation with these resources before handing over any evidence. Find out if their process and analysis will consume the sample, leaving you with no more samples for a follow-up analysis.

While I am somewhat pessimistic that DNA alone will furnish any paradigm-shattering revelations, I believe it is important to try. If anything, the years of research that have gone into the Bigfoot phenomenon have shown us it is indeed a marathon, not a sprint. I believe that DNA will be a large piece of the puzzle. Not necessarily the piece that solves the puzzle, but a significant one nonetheless.

When it comes to DNA evidence, it can be a bit more complex and involved than just seeing an item on the ground and picking it up and putting it in an evidence bag. Since DNA is often contained in bodily fluids, you may find yourself in a situation in which you are not only trying to collect dried liquid such as blood or saliva, but you also face the added obstacle of keeping your own DNA out of the mix and avoiding cross-contamination.

To collect something like dried blood, saliva, or other possible DNA-containing liquids, the most used technique at this point in time is a sterile swab. You should already have swabs in your evidence collection kit from the previous chapter. Medical grade, sterile swabs can be obtained at most box stores and internet retailers for a nominal cost. I suggest buying a box and keeping a few extras handy.

One challenge we ran into when collecting potential DNA evidence was when we would encounter scenes hours or even days old, and there was dried blood. If you rub a drop of dried blood with a swab, you may not get any transfer from the blood. Therefore, your best collection method is to rehydrate the blood by getting the swab damp. For this, it is best to use distilled water. Distilled water is a good hydration base because it has had the organic and inorganic components removed, posing less risk for contamination.

You can purchase a small dropper bottle of distilled water from most evidence supply stores as well as small plastic vials containing distilled water. These are marketed for the sole purpose of hydrating your swabs.

When the time comes to swab your possible dried blood, place a few drops on your swab. Avoid over saturating your swab. You do not want to submerge your swab in the water; just put a few drops on it. Then lightly contact the swab with the area you intend to swab and lightly twirl the swab around while touching the surface you wish to swab. Try to get good 360-degree coverage of your swab with the surface area.

After you feel you have adequately swabbed the area, set your swab in a safe place and let it dry for a minute or two before packaging it. We would use one method to put the swab back in its packaging with the swab sticking out the opened end and let it set on the packaging or prop it up and let it dry. Then we would tape the packaging closed around the swab when it was finished drying. Be especially cognizant of elements like the wind and other factors that could potentially disturb the drying process. If it is just not a permissible environment for drying, package the swab and quickly put it in a safe place. I like the swabs that come with their own plastic cap that can be secured over the swab.

DNA evidence does not necessarily have to be collected with a swab. You may find an item with potential DNA on it and decide to collect part of or the entire item. Like my industrious sergeant who cut up the vehicle's airbag to collect a bloodstain, you may find yourself in a situation where you have to think outside the box to collect your sample. For example, you may find yourself with an organic object, such as an apple or other item that you believe may have potential DNA evidence on it. It may be more practical to collect the item and swab it at another location. Further, if your testing facility is close by, it may be better to just take the entire apple to them and let them do the swabbing. Every investigation can be dynamic and fluid, and many different factors can influence how you can and should react.

DNA can be a great investigative tool. However, the high cost of getting samples analyzed can be off-putting to many citizen scientists and investigators. My advice is, if you plan on looking for DNA evidence, do a bit of homework beforehand and find some options available to you for potential analysis of anything you collect. Further, do not put all your investigative eggs into one basket; as someone who has put people in jail using

DNA evidence, I can attest that DNA is only part of the puzzle. You need a lot of documentation behind your evidence. Where was it collected? Under what circumstances? What other evidence was found, seen, and collected? All these questions and more should be answered in your field report.

CHAPTER 14
INTERVIEWING WITNESSES

As stated previously, interviewing witnesses can arguably be described as the bread and butter of Bigfoot investigation. For a creature that is camera shy and leaves little to no physical evidence, speaking with those who have caught a glimpse of the elusive creature can be the only insight we can get. An articulate, attentive witness interview can be a veritable goldmine of information.

That being said, most Bigfoot enthusiasts have little to no training in interviewing witnesses. Interviewing is both a skill and an art. I have had, both, training and experience in interview and interrogation during my law enforcement career. These chapters will primarily deal with the interviewing aspects of the investigation. Rarely, if ever, will you find yourself in a situation in which you will need to interrogate a witness.

An interview is a free-flowing and unstructured, non-accusatory question and answer session between an interviewer and interviewee. An interrogation is typically conducted when the subject's guilt is relatively certain and is accusatory in nature. The goal of an interview is to obtain information; the goal of an interrogation is to elicit a confession of the truth that someone is concealing.

Unlike what you have seen on television, there are no foolproof, magic-like skills that I can teach or tell you about, that will turn you into a human lie detector. For example, I cannot tell you that if a subject you are interviewing looks up and to the right while recalling a detail, they are lying, or if they tap their fingers on the table while speaking with you, they are being dishonest. However, there are certain behaviors and indicators that I can inform you about that are often associated with someone who is being deceptive. But, again, these are not foolproof nor infallible. Instead, these tips and methods give you more tools to put in your investigator's toolbox. The overall theme of this entire book has been documentation. Use the techniques in this book to document all the evidence and information, put it together, and then evaluate and come to your conclusions.

Some investigators are better interviewers than others. I personally was always a bit of an introvert and so was not always the best at interviews. Over the years, I found the best way to improve interviewing was to do interviews. Like most skills, it takes lots of practice. After many intense interviews, I would find myself coming out of the interview room, walking back to my office, and sitting at my desk, almost exhausted. Being completely absorbed in an interview and having all your senses focused can be mentally and physically exhausting. Do not be surprised if you come out of a particularly long interview feeling like you need a nap!

One of the first steps to conducting an interview is often one of the most overlooked, and that is, setting up the interview. Maybe your witness contacted you to tell you about their encounter; perhaps you heard about their encounter second or third hand and want to interview them about it. However you contact your witness, be polite and cordial with your communication with them. Semantics can often be your friend. When speaking with the witness, you may want to avoid intimidating words like "interview" or "investigator" and instead use the phrase "talk to you" and "enthusiast."

If your witness seems wary or unsure of partaking in an interview, address it, and ask them if there is anything in particular that is making them nervous about speaking with you. Address any concerns they have and try to be as accommodating as possible. Remember, they are taking time out of their busy lives to give you some of their time. Further, be sure to explain anonymity and privacy concerns. If you offer them complete anonymity, ensure that you do your very best to ensure it. Once you betray someone's trust, it can be near impossible to gain their cooperation in the future. Even if the information was somehow exposed accidentally, if it was entrusted to you, you should do everything within your power to maintain the wishes of your witness.

When scheduling the interview, if possible, conduct it in a comfortable place and with minimal distractions. For example, if the witness wants to conduct the interview at their residence, that is fine, but I would suggest ensuring the home is free of distractions. If the witness has several young children, the residence may not be the most ideal place to conduct the interview. Meeting in public is possible too, but make sure the meeting place is free of distractions. Sitting next to the children's play area at the local fast-food restaurant may not be the most distraction-free environment.

It may be necessary to conduct multiple meetings or interviews for one witness. Sometimes, a witness may want to meet in public initially and make sure you are not a threat before committing to an interview or future meeting. Try not to be offended by any such requests and try to be accommodating. It may help to have some examples of your work on social media or the internet that you can provide them to help expunge their fears.

Again, thinking outside the box may come in useful for scheduling interviews. Ideally, in-person interviews are the best, but other avenues should be pursued if scheduling is not working out. For example, a video chat interview or a written statement is better than no interview. Depending on your situation, you may have to resort to those or other alternatives to get your witnesses' encounter.

It can be challenging to determine how much time to allow for an interview. It would be best to schedule it in a perfect scenario when both parties have the maximum available time to give. However, this is often not the case, and one or both parties may have commitments and responsibilities to attend to. I like to tell my interviewees that I'd like to talk to them for about 45 minutes, but it could go a little longer. Rarely have I found I ever needed the entire 45 minutes, but it is better to have more time available and not use it, than to run out of time and need more.

Once you have your, hopefully, in-person interview scheduled on the books and you are ready to meet your witness face to face, let's talk about appearances. Some aspects of our appearance are out of our control. You may be almost seven feet tall and as wide as a barn door. That is not necessarily something you can change. However, if you are that tall and dress like an extra in a show about motorcycle gangs, you may give off an intimidating aura. Be cognizant of the clothes you wear to an interview. On the opposite side of the same spectrum, wearing a suit and tie may be intimidating to a witness, giving the appearance of a government agent or the police. When selecting attire for an interview, try to pick out neutral professional attire. A nice polo and khaki pants can give a professional appearance and be comfortable. Avoid shirts and clothing with risqué or offensive graphics and shirts that express a political or religious leaning. Wearing the clothing of a rival politician or party that the witness opposes can put the witness on the offensive and lessen their level of cooperation with the interview. Remember, you are a professional; you are there to obtain and document information. You are not there to convert the witness to your religion or political party.

When it comes time to conduct the interview, try to make the witness as comfortable as possible. If meeting in public, it may be beyond your control to furnish comfortable chairs or regulate the temperature in the room, but do your best. Having a comfortable witness makes it easier to get a baseline of their normal behavior.

At the beginning of the interview, be sure and walk the witness through the process. If privacy is an issue for the witness, reiterate any steps you are taking to ensure anonymity. Get permission from the witness before beginning any audio or video recording. I recommend audio recording the interview at the very least, if possible. I suggest having a notebook and pen in your hand or nearby during the interview. Explain to the witness that the notebook and pen are there so you can jot down a question you want to ask them or a detail you want to revisit without interrupting them while they are speaking.

You will want to begin the interview by talking to the witness about some subjects unrelated to the encounter you are about to ask them about. For example, talk about the weather, how long they have lived in the area, nearby family, hobbies, movies, music, and favorite foods. This "small talk" aims to build a rapport with the witness and establish a baseline. Pay particular attention while the witness is telling you about things that you know, for a fact, are true. Understanding how the witness conducts themselves when speaking about something true can assist you in noticing deviations from this behavior and help identify behavior indicative of deception later on.

Likewise, be aware of your own body language and behavior. Do your best to ensure there are no barriers between you and your interviewee. Barriers can have several different meanings when it comes to interview and interrogation. The most literal definition can refer to a physical barrier between you and the interviewee. Most likely, this will be a table or desk. If your interview location's space and floor plan allow it, try to sit across from your interviewee with nothing between you. This may not always be possible because of how the room is set up. Finding a place to seat two people with no table between them at a restaurant is likely impossible. There may be a need for a table or desk.

Perhaps you are sharing drinks or a meal, or the witness has brought items such as photos or other things they would like to share during the interview.

If the interviewee chooses to sit behind a barrier or not sit behind a barrier, pay attention to which they choose. Seated behind a barrier may signify that the interviewee is not entirely comfortable with your presence. It is not necessarily an indicator of deception but can give you a glimpse into their current mental state. Further, barriers such as desks and tables can obstruct your view of the interviewee and prevent you from seeing things such as crossed legs, leg shaking, and other physical manifestations of stress and/or nervousness.

Another barrier can be described as a closed-off body position. For example, arms and or legs crossed, physically turning the body in the chair to angle away from you. While keeping an eye out for these barriers, make a conscious effort that you are not displaying any of them. Sometimes, when an interviewee isn't telling you what you expect to hear or want to hear, you as an interviewer can start to display barriers and other indications that you are "checking out" of the interview. I advise against this and urge you to maintain an interested, open posture toward the interviewee.

As the interview progresses, you are likely to notice that the interviewee will literally "open up", arms and legs will come uncrossed, and they will lean forward in their seat to better engage with you. Pay attention to these changes in position and posture to get a better indication of their current level of comfort.

Some of the most difficult barriers you will face as an investigator are mental and societal barriers. Unfortunately, there are few effective ways to combat these barriers, as they are often difficult to identify. Most of the time, a witness affected by these barriers will never reach out or make themselves known to an investigator to even be courted for an interview. These barriers include fear of coming forward due to being ridiculed or other negative social stigmas that could be placed on a witness for "seeing or believing in Bigfoot." Many witnesses held back by these barriers will never even come to your attention. If you are approached or informed about such a witness concerned about these particular barriers, promises of anonymity are just about the only solution one can offer.

CHAPTER 15
INTERVIEW QUESTIONS

Now that the interview is scheduled, the venue is set, and you have sat down with your witness/interviewee, it is time to start asking questions.

I would give novice interviewers two pieces of advice to keep in mind above all else. The first is not to ask compound questions. Keep the questions short and simple. For example, do not ask questions like: “When you saw the creature, did you notice an odor, and what color was the hair?” More than likely, when asking those types of compound questions, you will get short answers for both questions that do not elaborate with more details, or you are only going to get one answer with details. The other question will go unanswered.

Second, and this is probably some of the most common advice you will hear in regard to police interviews, is to let them talk! Any interviewer who speaks more than the person they are interviewing is probably not doing a very good job. So when the witness starts to open up and talk, let them go. Do not interrupt. Only interrupt if the witness has deviated from the interview topic, and then only interrupt enough to get the witness back onto the topic. Remember your notepad? This is where it comes in handy. When the witness is talking, if they say something that you want to ask a follow-up question on, do not interrupt them to ask it. Instead, make a note in your notebook and let them continue with their story. When they are done and you are confident they have nothing to add, ask any follow-up questions from your notebook.

I like beginning the questions with a broad over-arching question. For example, "What are we here to talk about today?" or "Tell me about what happened at (insert location) on (insert date)?" Since each interviewee is different, you will often get different levels of responses. Some better orators may launch into their story telling a very detailed account that needs almost no follow-up questioning. Other interviewees may give short, curt answers that make information gathering seem like pulling teeth. Others still will be in the middle of the two and provide some details while making follow-up questions necessary for others.

Do your best to keep follow-up questions short and sweet. As stated before, avoid compound questions and only ask one detail at a time. Instead, try to ask open-ended questions and do your best to not insert the answer to the question in the question. For example, "Did the creature you saw have shaggy red hair?" Instead, try to phrase your questions more like, "What can you tell me about the creature's hair?"

If you insert descriptive phrases in the questions, like the aforementioned shaggy red hair, it can influence the interviewee into answering what they think you want to hear. The purpose of an interview is to obtain information, not to plant it. So, it is best to let the witness describe the events to the best of their ability and not give them possibly biased options.

Asking open-ended questions can be a difficult skill to master. It is not something that most of us practice in our everyday lives. Going into an interview, especially where you are not familiar with the witnesses' encounter, can prove challenging to establish open-ended questions. If you do not know what the witness will say, you have to develop your questions during the actual interview. This is one of the aspects that makes interviewing so difficult. Not only do you have to actively be engaged in listening, but you also have to be thinking ahead about what questions to ask next and how to articulate them in an open-ended manner.

Keep in mind that not everyone thinks the way you do. Some people are more detailed-oriented than others, which may affect how they respond to your interview questions. For example, you could ask one person if the deer they saw had any horns, and they will answer no. However, someone else could ask the same person if the deer they saw had antlers, and they would answer yes. One could argue that the difference between antlers and horns is just semantics. Both refer to hard, pointy growths coming out of an animal's head. However, a detailed oriented person may argue that deer do not, in fact, have horns; instead, they are equipped with antlers.

With that in mind, if you feel that your interviewee is not "getting" a question or perhaps answering it in a way you did not expect, try to reword the question and ask it again. There is nothing wrong with clarifying a question, either. Do not be afraid to tactfully ask the interviewee to explain themselves or go into more detail. If the interviewee uses a word or phrase you do not understand, ask them to clarify.

The most important aspect of interviewing is walking away at the end of the interview with the information the witness conveyed to you and that you understand what the witness has relayed to you. Remember, it's all about documentation. In the next chapter, we will discuss some psychological aspects of interviewing someone who has experienced a potentially traumatic experience, like observing a Bigfoot.

CHAPTER 16
TRAUMA-INFORMED INTERVIEWING

Every couple of years or so in law enforcement, some new tool or technique comes along that is regaled as the latest and greatest thing ever. A few years ago, I became aware of something called Trauma-Informed Interviewing. I was fortunate enough to attend a class, that included this new interview method and was even taught by one of the actual doctors that were instrumental in developing the technique.

I was sitting in the class, listening to Dr. Doe (not his real name) talk about how trauma can affect the hippocampus and how that can cause problems with memory recall. I visibly perked up when I heard him mention this. Having previously served in the military, knowing a bit about PTSD, and having had training in responding to officer-involved shootings, I realized I had heard this before. Many new methods, techniques, and tools that are "new" to one discipline are often brought over from other genres or disciplines after some enterprising thinker realizes how they can be utilized for more than just one thing. For example, much like how we are taking police and forensic work and applying it to the search for Bigfoot.

Having some knowledge of what Dr. Doe was talking about, I commented as he walked past my table.

“I know what you’re talking about. It’s like your memory is a puzzle, and after a traumatic event, the pieces get knocked on the floor, and you have to reassemble them.”, I said, trying to impress Dr. Doe with my knowledge of the subject.

He replied, “No! You’re wrong, and here in a minute, I’m going to explain why you’re wrong and show the science behind how it all works.”

Needless to say, I was a bit confused by his reply. Could science have explored even deeper into how trauma affects the brain? Was Dr. Doe about to unleash the newest, mind-blowing discoveries that were not yet public knowledge? Was Dr. Doe privy to information that the U.S. Military didn’t know? Humbled by Dr.

Doe's admonishment of my metaphor, I paid even closer attention, scribbling notes and preparing to learn the real science behind trauma and memory.

Dr. Doe droned on for some time, using many science and anatomy terms that were likely lost on a room full of cops. Finally, toward the end of his lecture, he said he would break it down for us, so everyone could understand. What follows is pretty much verbatim what he said.

"So, you're walking around holding a box full of photos, and those photos are your memories, and they are in chronological order. After a traumatic event, it's like you drop the box of pictures on the floor, and they get all jumbled, and you have to put the pictures back in order."

I still remember the flush of color to my face as I turned to the person sitting next to me and said, loud enough for the class to hear, "Didn't I just [expletive deleted] say that?!"

"Yeah…but you don't have doctor in front of your name," was my neighbor's reply.

Other than learning that pictures are a better metaphor for memory than a puzzle, I walked out of that class with the idea that trauma-informed interviewing could be an effective tool in Bigfoot investigations. In essence, experiencing something traumatic can affect how a person recalls information related to the event. Many of the ways this can manifest are similar to post-traumatic stress disorder.

Obviously, the sighting of a creature like Bigfoot has the potential to be very traumatic, depending on the circumstances and the observer. Even an innocuous roadside crossing could be a potentially traumatic, paradigm-shifting moment for a person with strong opinions of a world that does not have room for the existence of such a creature.

Trauma-Informed Interviewing is one of the reasons that I advocate for the open-ended questions. This form of interviewing relies heavily on open-ended questions.

These questions can be easier for a person with a trauma-affected memory to answer. Further, I suggest getting some business cards from local resources that deal in the therapy and treatment of PTSD. Put a couple of the cards in your notebook or wallet and if you come across a witness who appears to have been negatively affected by the event, give them a card and let them know they have the option to speak to someone about their traumatic experience.

Be cognizant of your interviewee's emotional state throughout the interview; ask if they need a break and have tissues or a napkin nearby for potential tears. Many researchers, enthusiasts, and investigators are at least reasonably certain Bigfoot exists. However, for someone who has never given the subject much thought or is a fanatical skeptic on the subject, an experience can be a dramatic event requiring a lot of processing. For some, it is the equivalent of going from the normal, everyday world to a world inhabited by literal monsters.

CHAPTER 17
INDICATORS OF DECEPTION DURING INTERVIEWS

As stated previously, I cannot tell you any magic tricks that will instantly alert you that someone is lying to you. However, I can make you aware of some possible indicators of deception. I cannot stress enough that these are not, in any way, positive proof that someone is being deceptive. Instead, they are tools to put in your interview toolbox and note in your notebook, if observed, during an interview. I recommend studying these indicators before an interview and developing a shorthand for each behavior, so you can document it in your notes.

I strategically placed this chapter after the chapter on Trauma-Informed Interviewing. Keep in mind that there are intersects between symptoms of PTSD and some of the indicators of deception. Undoubtedly, if you perform enough interviews, you will eventually interview a truthful person who will display or act in a way during an interview that will cause you to doubt their veracity. Inversely, you will undoubtedly interview someone lying to you that you will absolutely believe.

It's hard to say why precisely people lie. Generally, in police work, the reason was so that they, or their loved ones, would not get charged with a crime and taken to jail. In the world of Bigfoot investigation, the motivations get a bit more complicated. In some instances, people may lie because they feel they will somehow make money off of some form of Bigfoot evidence or encounter. Others may lie because they obtain a feeling of superiority over those they deceive with their tales. Some may sincerely believe they saw something unexplainable but simply misidentified a known animal. Others are not lying but relaying things they may have seen when having a break from reality due to a mental illness. Still, others lie because they want attention and notoriety from the community.

Those that lie for attention will often, eventually, display red flags. Anyone who has had multiple sightings, separated by a long period of time and distance, is likely not being truthful. For a creature as elusive as Bigfoot, it amazes me how many people just seem to stumble into them wherever they go.

The experience and training in interviewing I have had can often be a blessing and a curse. I cannot count the times I have been listening to an interview on a podcast or video, and by the end of the interview, I have completely discounted the witnesses' stories. Often, the witness will make a contradictory statement to something they have previously stated in the interview, and the host or interviewer either ignores it or does not catch it. This infuriates me to no end; was it an honest mistake? Or an indicator of deception? A few follow-up questions would usually clear such a matter up, but most podcasters are interviewing for entertainment, not the truth.

I remember one particularly thrilling interview I listened to on a podcast, and the witness stated that a person had died at the hands of a Bigfoot, and the local Fish and Game had covered it up and labeled it a bear attack. The witness did not realize that bear attacks are very well documented and easily verified. The witness stated this occurred in a state with a tiny, if any, bear population. I figured it would not be hard to verify the existence of this alleged coverup. No one was as surprised as I to learn that there had been no bear attacks reported in that state for over 100 years!

Were the conspirators so inept they forgot to report the cover story? Armed with this information, plus a few other indicators of deception I had noticed in the interview, I contacted the host and let them know that the witness was likely being less than truthful. The only reply I got was a, "Ummm…could be." It took me a bit to realize that some people, especially those who make money from the subject, are more interested in entertainment than facts.

More indicators of deception were found on a different podcast, coming from the host themselves. I listened to a Bigfoot podcast and perked up when I heard the host speak of a woman who lived in Alaska and her husband worked in the oil fields in the northern part of the state. The husband would leave the property frequently for his two weeks on, two weeks off rotation, and while he was gone, the woman would be harassed by Bigfoot on their remote property.

Obviously, living in Alaska, this very much caught my attention. As luck would have it, I ran into the host of the show, in person, at a convention, in another state, not long after the host spoke of the woman's plight on the show. I thought if I could convince the podcast host to tell me the nearest major city to the woman's property, I could use my investigative skills to locate her and offer assistance and possibly, an excellent research opportunity.

I introduced myself and told the host I was a law enforcement officer and that with my resources, I could potentially help the woman in Alaska he was talking about on the show. The host said he had no idea what I was talking about. When I began to add more details and wouldn't let the subject go, the host said I must have been mistaken, and he did not think it was Alaska. I assured him it was. I was from Alaska, and I would not have misheard something so exciting and potentially near me.

After speaking with the host for a few minutes, it became clear they were not going to relent and admit they had told the story I had listened to multiple times, and had saved on my podcast player on my phone. I expected the host to potentially tell me to pound sand and that they would never reveal the woman's location or identity due to requests for privacy. I have to admit, I was a bit taken aback by the denial of the story ever-existing and then the accusations that I "heard it wrong." I left the podcast host my card, and asked that they please check into it and ask the woman, who I was confident at this point, did not exist, to contact me. Needless to say, I never heard from the host or the woman.

You do not have to have any training to detect the deception indicators in the previous story. First, denying they said what they said, even though it was recorded. Secondly, accusing the other person of misremembering or mishearing what was said. All classic signs of deception and gaslighting. Interesting tactics to employ when a simple, "Sorry, I'm sworn to secrecy," would have sufficed.

Frequent deceptive behavior indicators can manifest through both verbal and physical means. Observation of one or more of these indicators is not definitive proof that the subject is being dishonest. However, these behaviors have been cataloged by interviewers and body language experts as being displayed by individuals who were deceptive and recorded in multiple interviews of subjects that were known to be relaying deceptive information at the time.

In person, interviews are best as they allow the interviewer to observe the subject's physical actions during the interview and observe potential deceptive behavior indicators.

Some behavioral cues exhibited by deceptive subjects can include such behaviors as:

- Retreating from the interviewer, placing barriers between subject and interviewer (crossing arms, legs).
- Frozen in their seat, with little to no movement.
- Bladed position, turning body away from the interviewer.
- Head and body slump.
- Touching face, pulling on nose, ears.
- Licking lips.
- Nervous movements, significant posture changes.
- Grooming behaviors include adjusting hair, clothing and picking at lint.
- Hiding of hands or feet (sitting on hands, for example).
- Leg bouncing and or finger taping, knuckle popping.
- Sweating, wiping sweat.

These are some indicators of deception that can manifest physically. Of course, there are more, but I have listed some of the most common here in the interest of brevity.

Another element of the interview process to keep in mind is eye contact. An average amount of eye contact is made between 30 and 60 percent of the time between two persons having a conversation.

The commonly heard, "Look me in the eye and tell me that," seems to have some veracity in its roots. Truthful individuals tend to make and maintain eye contact when making an honest statement. Deceptive individuals will tend to look away when making a deceptive statement. Granted, there are certainly exceptions to this rule. Some deceptive individuals may be aware of this and overcompensate and make an uncomfortable amount of eye contact during the conversation to seem sincere.

Keep in mind, especially dealing with other cultures, that there are different mindsets about eye contact worldwide. For example, in some cultures, it is considered rude to maintain eye contact, while in others, the level of eye contact given or received may be based on the person's age, gender, social status, or other factors.

The beginning of the interview, when establishing rapport, is an excellent time to pay attention to eye contact. When asked a question requiring a factual answer from memory, pay attention to where the subject looks. Typically, the direction a person looks when thinking of a response can reflect which hemisphere of the brain they are accessing to retrieve the information. If they look up and to their left, they are pulling from their memory. If they look up and to their right, they are creating an answer or accessing their creative mind to lie.

Everyone is different. There are some excellent liars out there and some terrible liars out there. Someday, there could possibly be technology available that fits in our pockets or smartphones that eliminate all doubt about whether a person is being truthful. Unfortunately, as of this writing, it is not here yet.

Some of the best liars I have encountered were those that never changed their stories. Even when confronted with irrefutable proof that they were lying, they would stick by their story. No doubt, if you conduct enough interviews, you will eventually run into someone like this. There is also no doubt that if you conduct enough interviews, you will believe a liar and doubt an honest person.

Some verbal indicators of deception can often be somewhat easier to pick up on, especially if you record your interviews, which you should. I usually would pick up on verbal signs of deception after the interview when I went back and reviewed the interview for my report. Do not beat yourself up if you do not catch everything the first go-round.

I've found that qualifying statements can often be a good indicator of deception. Examples of qualifying statements can include phrases like, "This really happened." Or "To the best of my knowledge." Along with these, I urge you to keep an ear out for unnecessary boasting.

Unnecessary boasting is one of my pet peeves during an interview. Generally, I try to give everyone the benefit of the doubt and believe them unless they give me a reason not to. Once someone starts going on a tirade about how honest and trustworthy they are, I automatically begin paying closer attention to looking for other signs of deception. I can recall two interviews where the suspect literally pulled out a Bible to show how much of a Christian they were and that they did not lie. Both suspects were lying during the interview.

It is not necessarily a sign of deception because someone tells you they are an honest, God-fearing, hardworking person at the beginning of an interview. However, repeated, prolonged exposition on their veracity should start to raise red flags for any interviewer. If I were about to sit down and tell someone a story that I knew sounded unbelievable, I would probably mention how I was a law enforcement officer for many years, and that integrity was very important to me. Nothing wrong with a statement to establish credibility. It is the interviewees that will turn the bulk of the interview into establishing their credibility that should become the subject of more scrutiny.

Other indicators can include answering a question with a question or restating a question before answering. These are typical stall tactics that are used to give the interviewee time to formulate an answer they believe will satisfy the interviewer.

Changing their statement while telling the story can be another indicator to keep an eye out for. Again, I'm not referring to misspeaking and correcting themselves, but more like changing the events or actions that occurred. Getting the timeline of events confused is one thing, normal even when dealing with a traumatic experience, but a significant change in events can be an indicator of fabrication.

Regardless if you feel the person you are interviewing is not being honest with you, you should try to end every interview by leaving the door open for follow-up contacts. Tell the interviewee you would like for them to contact you in the future if they think of anything else they would like to tell you. Further, tell them you would like to be able to reach out to them if you think of any more questions. Inevitably, you will think of a question or detail that you would like to introduce or revisit, and leaving the door open for further communications with your interviewee can make that much easier.

My last bit of advice for in-person interviews would be to tell you to practice. You do not necessarily need to interview someone about a Bigfoot sighting to practice an interview. In the Army, we had a thing called hip pocket training, in which, when faced with downtime, the sergeant would pull out his manual for common tasks out and start quizzing or instructing us on a particular topic. Interviewing can be an excellent skill to practice impromptu. You can interview a co-worker about an incident at work or what they did that weekend. You can interview a friend about a place they went to or a vacation they took. Practice your open-ended questions and watch body language. It will help you.

CHAPTER 18
WRITTEN STATEMENTS

During my law enforcement career, I was fortunate enough to attend a couple of classes where the main topic was written statements. I came out of these sessions a pretty big proponent of written statements and how they can be used in a criminal investigation. However, most law enforcement agencies seem to be moving away from obtaining written statements and focusing more on in-person statements and interviews.

Written statements can be an excellent resource for obtaining a statement from the witness to an event or sighting. Further, they can offer a look into the truthfulness of the statement. Obviously, factors can come into play, such as how well educated the statement writer is and how much of a grasp they have of the English language. These factors should be taken into consideration before asking for a written statement.

There is nothing inherently wrong in asking a witness for both a written statement and an in-person interview, although most will opt for one. Both can take a lot of the witnesses' time, and most investigators will tell you the in-person interview is best. Despite being a fan of written statements, I believe in-person interviews are better. Written statements are good for when in-person interviews are not likely to happen and attempts to meet in person have not worked out.

During the class where trauma-informed interviewing was discussed, a detective related a challenging case to the class. The detective opined that a female in his jurisdiction was mentally ill and had reported a sexual assault, but when it came time to interview the victim, she would not disclose what had occurred. The detective said the victim was just not comfortable speaking to the investigators about what happened. After several attempted interviews, his investigative team felt as if they had exhausted their options for helping the woman.

I overheard the detective sharing his frustration with the class and chimed in, asking if they had asked the victim to write out a written statement. I could literally see the lightbulb go off in his head when he said, “No…we haven’t thought of that.”

I do not know what happened with that case or if he went back home to his own jurisdiction and attempted to get the victim to write her experiences down instead of talking about them. However, I often think about it and hope that it worked and that some progress was made in the case.

For this chapter, I will stick to written statements and how their structure can relate to the statement’s veracity. However, other interesting factors can be discovered by examining written statements. I suggest if this subject fascinates you, to explore it further. There are plenty of materials and classes available to civilians for you to peruse.

When asking your witness for a written statement, you will want to ask them to begin their account of the incident sometime before the incident. For example, if the incident of interest occurred at 4PM in the afternoon, you may want to ask them to begin their written statement at lunch. Ask your witness to write in as much detail as possible, everything they can remember starting at lunch, up to, and including the event, as well as after.

When you receive the written statement from the witness, it is important to read the entire statement. Pay particular attention to how much detail the witness uses to describe things unrelated to the incident. Below is an example of a good, detailed, written statement:

Billy and I finished setting up camp around lunchtime. I cooked burgers in the RV, and I accidentally cooked them too long, and they were a bit crispy and burnt, but we ate them anyway. Billy had brought some sodas in the cooler, and we had them with our lunch. We put our trash in a bag, and I walked about fifty yards out of camp and was going to tie the bag in a tree to keep it away from bears. I had a rope tied around one end of the bag, and I was going to toss it over a high limb and tie it off so the bag would be out of the way. I tried to toss the rope up over a limb two or three times and had missed. So I began to tie a knot in the end of the rope to make it heavier to get it over the limb. I was looking down and tying the knot when I heard a noise behind me. I don't remember the noise; it sounded like leaves crunching or something. I felt there was a person or animal behind me, so I turned and looked.

I saw a giant hairy man looking at me a few yards away. He was about 7 ft. tall. I know this because, later, Billy and I went back, and I had Billy stand where the hairy man stood. Billy is six feet tall, and this thing was at least a foot taller. I marked the tree and took photos. The hairy man-thing had reddish, brown hair; it was standing in the shadow of a tree, so it was hard to tell for sure what color it was, but that is the best description I can give of its hair. There was no hair on its face, and the skin on its face was kind of ashy gray. I didn't see the whites to the eyes, just big black eyes. Its face looked kind of like a neanderthal, and it didn't make any noise. It was standing by a tree, and the left arm was just hanging down by its side with the hand down by his thigh kind of in a cupped shape; the other arm was stretched behind the tree. I don't remember seeing the right hand at all. I didn't notice any odors other than what was coming from the garbage bag and couldn't' see the hairy man's feet because the foliage was up to about his knee.

I think it was looking at the garbage sack I was holding. It didn't move or make any noise; it just stared. I don't think it ever blinked while I was looking at it. I probably only saw it for twenty seconds or so. It felt like more, though.

I got really scared, dropped the bag and rope, ran back to camp, and told Billy. We went back about a half-hour later, after I worked up the nerve to go back. Billy brought his rifle, and I had my pistol. It was not there anymore, and my garbage bag and rope were gone. Billy walked around the tree where I saw it and found where the plants were kind of mushed down. I was too scared to walk around and talked him into tearing down camp and going home. We went back a few days later and took the photos with Billy next to the tree. We didn't see or hear anything strange then.

The previous statement is a decent example of a detailed statement. If you receive a statement similar to this, I would suggest you first read through the entire statement. Next, you should separate the statement by subject; you may want to make a copy of the statement for this. Finally, take two different color highlighters and choose one color for the subject for the statement, the sighting, and one color that is for parts of the statement that are unrelated to the incident, like cooking lunch, dealing with the trash bag and setting up camp.

After using your highlighter to separate the relevant subjects from the non-relevant subjects, you can physically see how much of the statement the sighting or incident composes using the two separate colors. This can give you a good sense of scale pertaining to the statement. For example, even though the witness stated the sighting only lasted twenty seconds, he wrote just as much, if not more, about it than he did about cooking lunch and setting up camp. No doubt, cooking lunch and setting up camp took much longer than twenty seconds.

This is not proof that the sighting actually occurred, but it is a good exercise to conduct with written statements. Beware any written statement in which the subject of the statement takes up the least amount of volume in a statement. Below, I will give you an example, using the same encounter, of a poorly written statement.

Billy and I finished setting up camp around lunchtime. We cooked burgers on the campfire, and I accidentally cooked them too long, and they were a bit crispy and burnt, but we ate them anyway. Billy had brought some sodas in the cooler, and we had them with our lunch. We put our trash in a bag, and I walked about fifty yards out of camp and was going to tie the bag in a tree to keep it away from bears. I had a rope tied around one end of the bag, and I was going to toss it over a high limb and tie it off so the bag would be out of the way. I tried to toss the rope up over a limb two or three times and had missed. So I began to tie a knot in the end of the rope to make it heavier to get it over the limb. I was looking down and tying the knot when I heard a noise behind me. I got the feeling I was being watched and turned around.

I saw a Bigfoot just a few feet away from me, standing by a tree. I got scared and ran away. Billy and I came back later and didn't see anything. I swear this really happened. I worked for the city for thirty years, and they will all tell you I'm an honest person. So I have no reason to lie about any of this.

As you can see by this version of the statement, the witness spends a reasonable amount of time setting up the sighting and giving details of what happened before the sighting. However, the actual sighting and aftermath only comprise a few sentences. This is the type of statement that should invite extra scrutiny and skepticism.

Again, this does not mean that the sighting did not occur. Ideally, you will take other evidence and observations from your investigation, document them, and then you may have enough information to reach a conclusion on the validity of the report. This method of statement analysis is just another tool to put in your toolbox.

CHAPTER 19
REPORTS

You have taken the sighting report, talked to the witnesses, gone to the scene and investigated, collected any evidence you found and took photos; now what? Ideally, you will want to combine all this information into a report. I find reports a good way to combine the data collected and articulate the steps taken in an investigation. Often, I would look back at reports written and discover that I had forgotten specific details that I had, luckily, documented in the report. With cases going to trial months and even years after they occur, the act of documentation and just writing down what happened should not be underappreciated. The small detail you write down today may help you find parallels between cases you investigate later.

At a homicide class, I attended out of state, one of the instructors was from the local large metropolitan police department. She kept mentioning investigative letters during her class. I had never heard of an investigative letter before or since that class. I sat there silently, wondering what the heck an investigative letter was and whether I should ask or not because no one else in the class seemed to be confused by the term. Finally, during the class, she handed out a stack of materials that included examples of investigative letters that detectives in her agency had written. I have to say, I didn't know what to think when I started reading the mysterious investigative letters only to find out they were just reports.

I never found out why the agency played the semantics game with their reports and called them investigative letters. I imagine it probably has something to do with some local court rule that somehow makes the police department and local prosecutor's lives easier by calling it an investigative letter rather than a report.

I tell you that so you can see that all police departments do not use the same terminology, and they are not all necessarily on the same page. So all I had to do was look at one of these mysterious investigative letters to see that it was actually something I was very familiar with, a report.

Imagine how awesome it would be if all Bigfoot investigators spoke the same "language" and structured our investigations similarly? After the tragic events of September 11th, 2001, many law enforcement agencies around the country found that they used different codes and terminology for their radio traffic and began to make an effort to move toward "plain speak" on their radios so that they could better communicate with other agencies during a large-scale event. Wouldn't it be great if all Bigfoot researchers and investigators were on the same page?

Ultimately, the work you do toward investigating the Bigfoot phenomenon is yours. You put your time and effort into it for motivations only known to you. Maybe you just want to know what Bigfoot is or if it exists? Perhaps you want to be the one to prove it to the world and enjoy the fame and riches that might come with such a distinction? Regardless of your motivations, organizing and compiling your findings into a format that is easy to understand and you can easily share with others, will be an important part of any research or investigation.

Remember when I was talking about evidence collection, and I said that when diagraming and photographing a scene, anyone should be able to come along, at any point in time, look at the diagram, and reconstruct what occurred? The same goes for your reports and case files. Sometimes, the person looking over the work years later might just be you. Sometimes, it may be someone else entirely.

Many years ago, I arrested a young woman for driving under the influence of alcohol. It was a pretty routine traffic stop and arrest; I had forgotten about it over the years. The memories of the stop were eventually overshadowed and replaced by hundreds of other arrests over the years.

Five or six years later, I got a message from a sheriff in the lower 48, (that's the contiguous states that are not Alaska and Hawaii, for you non-Alaskans), requesting that I call him. I had no idea what a sheriff in another state might want with me and was quite curious about how this call would go.

I called the sheriff back and was surprised to hear that the young woman I had arrested so many years ago had been involved in another drinking and driving incident. Only this time, someone had lost their life. The young lady was being charged with murder, and the sheriff and prosecutor wanted my report for possible use in her prosecution. I quickly obliged the other agencies' records request and got them the copies of my report and other paperwork they needed.

When I was writing that simple little DUI report five or six years prior, it never once entered my mind that the report I was writing might be in the hands of a sheriff or prosecutor from another state for possible use in a murder trial. But, you never know who will be looking over your work someday and how it may influence a bigger, broader subject; so do good work!

Hopefully, I have articulated how important good documentation and organization can be. That being said, just how do you write a report?

There is no universally accepted template for how to compose an investigative report. However, I do have a particular format that I like to use. As a general rule, you will want the who, what, when, where, and how of whatever you are writing about to go into your documenting efforts. Although sometimes, you may not have all of that information, that is not necessarily a problem; however, it is good practice to document why you do not have that information.

The following pages will be an example of a report that I have put together. This is an example report. While it comprises an amalgamation of different sightings and investigations, I have taken liberties with the sighting and investigation to add as many examples into one report as I could. Therefore, it will not be uncommon for reports to have less information than this report. For example, some investigations will be incidents reported months or even years after the fact, and the location of the sighting could be long gone and is now the sight of a suburban neighborhood. In addition, some sighting locations may be closed to the public, and you cannot reach them. Factors such as these may rule out the possibility of a scene investigation. Further, some reports will, hopefully, produce evidence that will need to be documented in the report, while others will not.

Again, this is just an example; feel free to work with the format to produce a report structure that works for you and your team.

CASE #22-003

SYNOPSIS:

On September 18th, 2022, I was contacted via social media by John Doe. Doe reported he had a visual sighting of a Bigfoot on September 15th, 2022, at the Rainbow Lake Campground. Doe advised he had been camping with a friend, Bill G. Doe agreed to meet with me to discuss the sighting, and we arranged a meeting. Bill G. also spoke with me as well at a later date. My research partner and I went to the sighting location and found a track in the area that was cast and suspected hairs. The investigation is closed pending analysis of evidence.

INITIAL REPORT:

The following message was received at 1430 hours via social media to the Bigfoot Research Team message inbox.

John Doe: I saw your page and wanted to tell you about something that happened at Rainbow Lake Campground this last Thursday. My buddy Bill and I were camping, and we had just finished lunch. We were going to leave camp to go fishing, and I went to take our trash from lunch away from the campsite to tie in a tree to keep it away from bears. A few yards away from the camp, I saw what I can only describe as a Bigfoot watching me. It scared the hell out of me, and I ran away. Later, Bill and I came back and measured a tree that I saw the creature next to. It had to be almost seven ft. tall. I have pics of the tree and Bill standing next to it. The orange tape shows about where the head of the thing was. I just wanted to let someone know. I have been fishing and hunting all my life and have never seen anything like it. There was no way it was someone in a suit or a bear.

INVESTIGATOR'S ACTIONS:

I replied to Doe's message minutes after receiving it and asked him if he would be interested in meeting me to talk about his encounter. Doe responded, and a meeting was made for Monday, September 19th at 1200 hours at a local restaurant. Doe stated he could talk while on his lunch break from work. I agreed to meet Doe and speak with him and purchase his lunch. Screenshots of the messages received from Doe were saved in the case file in the photos file.

INTERVIEW WITH JOHN DOE ON 09/19/22: (RECORDED ON AUDIO FILE INCLUDED IN CASE FILE):

I arrived at the meeting place ten minutes early and procured a table in the quietest part of the restaurant. Doe arrived on time and sat with me at the table. Doe was dressed in his place of business uniform and requested I leave his employer's name out of all investigative reports and documents. I assured Doe this would not be a problem. Doe and I ordered our meal, and I acquired his permission to record our talk.

I began by asking Doe about how long he had lived in the area and his favorite places to hunt and fish. Doe articulated a good amount of information about local lakes, wildlife management areas, and campgrounds. Doe seemed to have a good understanding of the local wildlife laws and types of wildlife in the area, even showing me photos of some of the fish and game he had taken in the region over the last five or so years he had been living here.

During our talk, Doe seemed open and honest. I did not observe any overt indications of deception. After talking with Doe about himself and his affinity for the outdoors for several minutes, I asked Doe what we were there to talk about today.

"I saw Bigfoot." Was Doe's reply. I asked Doe to tell me about it.

In essence, Doe stated he and his friend Bill G. had planned to camp and fish at the Rainbow Lake Campground starting on Thursday the 15th and returning home late Sunday. The men drove to the campground in Bill's RV, arriving early in the afternoon around noon. Both men had left work early that day in order to secure a campsite of their liking.

Doe and Bill frequently camped together, both with each other and with their wives, but their wives did not attend this trip as it was a "boys' weekend." Doe and Bill planned on fishing most of the weekend and had brought a cooler with several different varieties of beer and spirits for their trip.

The two quickly set up the RV. Having camped so many times together, Doe stated he and Bill had the setup of the RV and campsite down to a science and could do it in minutes.

Doe stated the two men decided to have lunch and then would begin fishing. Doe said the "boys' weekend" tradition was that they did not start drinking alcohol until the first fish was caught. Bill had brought a few sodas and water bottles in one of the coolers for them to enjoy before the fishing and the drinking began.

Doe said he had cooked cheeseburgers on the stove in the RV for their lunch. Doe stated he had got distracted outside by watching an eagle and had accidentally burned their burgers. The two men ate the burgers anyway and had sodas with their burgers. The plan was that after lunch, they would launch their boat, and only after the first fish was hooked and in the boat would they begin to drink beer.

Doe was very adamant that I knew and understood that neither man had consumed any alcohol at this point. Further, Doe stated that whoever operated the boat typically would only have two beers until the fishing was over and would be "allowed" to drink more when they arrived back at camp. Doe said they had a good time but tried to be responsible. Doe said after lunch, he had collected the trash, grease, and other refuse made from their time in camp so far and had told Bill he was going to tie up the garbage. Doe said some years ago, he and Bill had an incident involving their trash and a bear at a different campsite and that ever since, Bill has been very strict about not keeping trash in or around the RV.

Doe stated that since he was the designated "camp cook", typically Bill would handle the trash, but there had been a problem with the boat that Bill was working on. So, Doe had decided to pitch in and take care of the trash to save time, so the pair could get to fishing.

Doe said he had taken the trash and the rope used to tie the trash up into the woods. Even though they had camped at that site before, Doe did not know where Bill usually tied the trash up. Doe walked into the woods in the direction he had seen Bill take the trash in the past, and walked about fifty yards into the woods and chose a random tree to tie the trash in.

Doe said he had trouble with the rope and the bag, he was having trouble getting the rope to go over the branch. So, he decided to tie a knot at the rope's end to make it heavier and, hopefully, get it over the intended branch. Doe said he had his head down, focused on the rope, and tied the knot when he heard a sound behind him. Doe could not describe the sound but said it was like the crunch of a footstep to the best of his recollection.

Doe stated he felt a presence behind him, like when you are in a room alone, and someone walks in behind you. Doe did not believe it was Bill because Bill was working on the boat, and he was sure Bill did not see what direction he went into the woods with the garbage. Doe said he turned around, expecting to see either a bear or a person.

At this point in the interview, Doe's demeanor changed noticeably. Before, while speaking about fishing and camping, Doe had seemed to be in a good, happy mood. However, when the topic of the actual sighting came into the conversation, Doe became noticeably uncomfortable and nervous. Doe leaned in closer when speaking about the incident and his voice lowered.

Doe stated that he saw what he could only describe as a giant man covered in hair when he turned. As soon as he laid eyes on it, Doe knew it was Bigfoot. Doe had heard of Bigfoot through TV shows and commercials but had never given it any thought. Doe's father had told him that there was no such thing as Bigfoot when he was young, and Doe had never really thought about it much since.

Doe said the thing had a face similar to a neanderthal, but the nose was flatter and wider. Further, he said it looked like it had reddish-brown hair everywhere he could see except for its face and hand. He estimated the hair to be between 4-6 inches long and described the face as "weathered and gray looking."

It was standing next to a tree and had one arm, the right one, extended out behind the tree. Doe said he had the impression the thing was holding on to the tree with its right hand but could not see it because of the thick foliage on that side of the tree. The left-arm hung, relaxed, and the hand was cupped near the creature's thigh. Doe said he thought the creature's arms were longer than an average person, but he wasn't sure.

He could not see the creature from the knee down due to the tall grass. Instead, the creature seemed to be staring intently at Doe's hands or the garbage bag and rope in his hands. Doe estimated he stared at the creature for about twenty seconds but admits that his sense of time was probably distorted due to the intense emotions of the moment.

Doe said the creature made no sound or movement during his observation of it but does recall noticing that its chest rose and fell as it breathed.

Doe described his initial reaction as confusion because he did not know what he was looking at; after the first few seconds, the confusion turned to fear. Doe said he had spent a lot of time in the woods, and while he respected the woods and the wildlife, he was not afraid of anything in the woods. Doe said he has had several encounters with bears in which he has always been able to scare the bear away, and he had never been harmed by any animal or situation he had encountered in the woods.

Doe stated that once the reality of the situation struck him, he became afraid. He stated he did not know what to do or how to react. Finally, his fight or flight response kicked in, and Doe said he dropped the rope and garbage bag and began to run back toward camp. He said he looked back several times, but once he took a few steps, his line of sight became obscured by trees and foliage, and he did not see the creature again.

Doe said he ran back to the camp and found Bill, working on the boat, almost in the same position he had last seen him in. Doe said he "panicked" and began yelling and packing up the campsite. Bill thought Doe had been chased out of the woods by a bear and retrieved his rifle from the RV. Doe stated that Bill did not understand what was going on, and he understood why, because at this point, Doe was not making sense and was in a blind panic with his only goal being to get away from there.

Bill got aggravated with Doe's panic and began yelling at him to calm down and tell him what the hell was going on. Doe realized that he could not leave the campground without Bill's cooperation because Bill had the keys to the RV, so Doe said that realization forced him to slow down and describe what had happened coherently.

At this point in the interview, our food arrived. As the waiter approached with our food, I noted that Doe stopped talking and did not resume talking about the incident until the waiter was out of earshot. Doe's actions appeared to reflect the reactions of someone who was possibly embarrassed by the stigma of being associated with a Bigfoot sighting. We took a short break to consume most of our food, and Doe continued his account a few minutes later. I noticed Doe check his watch when the food arrived, and he stated he had plenty of time until he had to be back to work.

Doe stated that once he could articulate what he had seen to Bill, Bill got excited and wanted to go back to the area and look for the Bigfoot. Doe had pleaded with Bill to just leave, but Bill said they would not until they returned to the area with their guns and camera.

Doe said that Bill watched Bigfoot TV shows every once in a while, and said they had to go back and do a recreation or try to get a picture if it was still there. Doe said he continued to protest and only agreed to go with Bill after Bill told him he was going into the woods with or without him. Doe got his pistol out of the RV and put it on his belt. Doe said he always brought the pistol camping with him. After the incident with the bear, he had worn the pistol religiously around camp but eventually became complacent and had started leaving it in the vehicle.

Armed with their firearms, camera phones, and some flagging tape, they slowly made their way back to the area. Doe said at this point, he was so angry at Bill for making him go back into the woods that he had told himself once they got back home, he would never talk to Bill again.

Upon arrival at the tree, Doe showed Bill the area, and both noticed the bag and the rope were gone. Doe pointed to the tree where he had seen the creature, and Bill ran over to the tree "like a kid on Christmas," Doe said. Doe said that Bill had told him the grass was all mashed down around the tree where Doe had pointed to and tried to get Doe to come over and look. Doe flat out refused and told Bill to take his pictures so they could get out of there.

Bill asked Doe to show him where the creature's head was, and Doe pointed it out and confirmed when Bill had held his hand up high enough. Next, Bill tied a small piece of flagger tape to a branch and asked Doe to take a couple of pictures. Doe complied and urged Bill to leave the area so they could go home.

Doe stated the entire time the duo was back in the woods, he had an overwhelming sense of dread. He thought the creature was going to jump out at any moment and attack them. Doe said he had even formed a plan that if there were any signs that the creature was still around, Doe would "do what he had to" to get the RV keys out of Bill's pocket and leave with or without Bill.

Eventually, the two made it back to the RV, and Bill tried to convince Doe to stay and continue the trip. Doe refused, and Bill relented, and they packed up and left the area.

Doe said he got home and spent the rest of the weekend "sleeping, drinking and looking up Bigfoot stuff on the internet." Eventually, he found the Bigfoot Research Team page and sent the message that preceded our meeting.

Doe then asked me several questions regarding if I had ever had a sighting and what I thought Bigfoot was. I answered Doe's questions and asked a few follow-up questions of my own.

I asked Doe if he felt that there was any chance that Bill or someone else had played a trick on him. Doe stated he was absolutely sure that no one was playing a joke on him. Bill had no way of knowing where he would go to tie the garbage off, and even if he did, he did not have the means and resources to produce a hoax of the quality that he saw. "It was a real animal, a live animal," Doe said. He was adamant that it was not a suit or a cardboard cutout.

"It was better than any Bigfoot movie suit or computer-created graphics I'd ever seen," Doe said. "It was either the real deal or a hallucination, and I've never had a hallucination before, so I don't know why I'd start now."

I asked Doe if there were any odors in the area, he could tell me about. Doe said that he did not smell anything except when he was tying the knot in the rope, and he had squeezed the garbage bag between his forearm and side. That movement had caused the bag to expel some air, and he had got a "face full of garbage odor, mostly burnt smell from the grease in the bag." Other than that, he recalled no odors from the incident or from when he and Bill had returned to the area.

I asked Doe if he thought Bill would speak with me about the incident, and he indicated he likely would. Doe stated Bill worked for a local government and would probably not want his name and info put into a report or online. I assured Doe that Bill could be anonymous, and Doe gave me Bill's contact info.

I asked Doe if he would accompany another researcher and me to the area and show us where the incident had occurred. Doe stated he would not be going back to the site but was happy to show me the area on a map. I brought up a map of the campground on my smartphone, and Doe pointed out the campsite and the place where the encounter occurred. He stated that the flagging tape was probably still on the tree and that with the directions he had given me and the pictures of the tree he had sent, I should be able to find the area with no problem.

Doe then excused himself, saying he had to be back at work soon. We shook hands, and Doe departed the restaurant while I paid the check.

WEATHER REPORT:

The weather report for September 15th, 2022, for the area near Rainbow Lake Campground was a high temp of 65 and a low of 40. Partly cloudy skies with periods of sun throughout the day.

INVESTIGATOR'S ACTIONS:

I reached out to Bill G. via the contact information Doe had given me. Unfortunately, I could not contact him directly and left a voicemail asking him to call me back.

I then reached out to Thom T., a member of the Bigfoot Research Team and my investigative partner. We planned to meet at the Rainbow Lake Campground on Tuesday, September 20th, and attempt to locate the area of Doe's sighting.

SCENE INVESTIGATION:

Thom and I arrived at the Rainbow Lake Campground at 1200 hours. The campsite Doe and Bill had been camped in was unoccupied. Thom and I parked our vehicles at the campsite.

Nothing of note was noticed in or around the immediate campsite area. It appeared to be a standard campsite that was regularly used. The ashes in the fire ring were cool, and the campsite did not appear to have been occupied the previous evening.

Thom noted that there seemed to be a lack of squirrel activity in the area and did not observe or hear any squirrels during our initial arrival at the campsite.

We walked into the woods in the direction Doe had indicated he had walked and walked back to the area Doe had shown on the map. My GPS tracked our route, and the coordinates and map data were downloaded into this case file. We found the tree with the marking tape on it and took several photos before approaching the tree. Upon approaching the marked tree, neither Thom nor I noted any pressed down grass or disturbance on the ground near the tree. Being taller than I, Thom stood under the tape, and I took a few photos of him in that position. The tape was measured to be about six feet, 10 inches off the ground.

I began to look around the general area where I suspected Doe had tried to tie the garbage in the tree. I was looking for any sign of the rope or garbage bag when Thom called me back over to the area of the tree. On the tree, about five feet from the ground, Thom discovered two (2) reddish-brown hairs about 4 inches in length. The hairs appeared to have been caught on the bark of the tree. The hairs were photographed, proper PPE was donned, and the hairs were collected separately and stored in individual evidence bags.

The tree's location was marked with a dropped pin in my GPS unit. Thom and I continued to look around the area and focused our search on the area behind the tree, walking to the northeast.

Forty yards to the northeast of the marked tree, we found an impression in the ground in a patch of dirt that was not covered in grass. The impression appeared to have five toes and the general appearance of a human-like footprint. Measurements were taken, and the footprint was 15 inches long and about 7 ½ inches wide at the widest point (see photos in the case file).

The decision was made to attempt to cast the track. Thom marked the track's location with a small orange flag, and I dropped a pin for the track on my GPS. Then, we returned to our vehicles to get more casting supplies.

We estimated our trip to the vehicles and back took around 20 minutes. When we arrived back at the track's location, both Thom and I noted an odor in the air. We both noted the odor at the same time. Thom stated it reminded him of the smell of a horse stable or barn. I advised Thom that I felt the scent was similar to the odor at a zoo.

After debating how to proceed for a few minutes, we decided to continue with the casting of the track. After pouring the plaster, we noted the odor had dissipated, and we did not smell it anymore. After the plaster was poured, we continued to walk around the general area and did not observe anything else of note.

We went back to the campsite and ate lunch, giving the plaster more time to set. We observed a squirrel come into the camp during our meal, looking for scraps.

We returned to the track and retrieved the cast. Again, no other odors or sounds were observed. We stayed in the area for another hour and then returned to our vehicles.

INVESTIGATOR'S ACTIONS:

Evidence was packaged and stored at my residence. Plans are being made to return to the area for an extended camping expedition.

INTERVIEW WITH BILL G. (VIA TELEPHONE, RECORDED):

On Wednesday, September 21st, I received a call from Bill G. (not his actual name). Bill stated he was out of state at the time and was not available for an in-person interview. However, Bill said he was willing to speak with me over the phone, and he gave his permission for the call to be recorded. Bill requested that his real name and place of employment not be used in any investigative notes or reports.

I asked Bill if he could tell me about the incident at the Rainbow Lake Campground on September 15th. In essence, Bill gave the details of the story that Doe had, including the burning of their lunch burgers and the problem with the boat, that Bill was trying to fix.

Bill stated he was working on the boat and was almost done when Doe had come running out of the woods.

Bill had heard Doe running through the woods before he saw him arrive at the campsite and had already begun to move toward the RV to retrieve his rifle because he was not sure if the noise was Doe or a bear. However, he saw Doe come bursting out of the trees and said Doe immediately began packing his belongings up and throwing them in the RV while looking over his shoulder toward the wood line.

Bill said that Doe kept telling him, "Let's go, let's go, we gotta go!" Bill thought that Doe was acting irrationally and did not think that he was acting that way because of a bear. Bill stated that he and Doe had been in all sorts of situations with bears, and he had never seen Doe lose his composure like he was now.

Eventually, Bill said he was able to calm Doe down and get him to tell him what had occurred. Bill said he got excited because he did not think Doe was lying or mistaken. In addition, he was familiar with Bigfoot and followed the Bigfoot Research Team's social media under a pseudonym.

Bill eventually convinced Doe to return to the area with him. Bill said he could tell that Doe was "not happy" with him or the idea of going back. Bill said there was no way he wanted to leave without getting a chance to see what Doe had seen.

Bill said they had gone back to the area and described finding the mashed-down vegetation near the tree and taking the photos with the marking tape for scale. Bill said that Doe was a "nervous wreck," he began to feel sorry for him and finally agreed to leave the area and take Doe home.

I asked Bill if there was anything else he could think of about the trip, and he said there was not. Bill stated he would contact me if he thought of anything else and asked if he could tag along if the Bigfoot Research Group went back to the area. I told Bill we would be in touch with him when he got back in the state.

EVIDENCE COLLECTED IN THIS INVESTIGATION:

23-003 Item #1 Single hair recovered from the tree at the scene.
23-003 Item #2 Single hair recovered from the tree at the scene.
23-003 Item #3 Cast track recovered from Lat/Long (123,456,23).
147 Digital photos stored in the case file.
3 Digital audio files are stored in the case file.

STATUS OF EVIDENCE:

Items #1 and #2 were mailed via USPS to Enogene Labs for DNA testing. Tracking #12340987565. Results of the analysis will be added as a supplemental report when received.

Item #3 was dried, cleaned, and photographed. Photos are saved in the photos section of the case file. Item #3 is currently being stored in a secure, safe area at the investigator's residence.

CASE STATUS:

Initial investigation into reported sighting is closed pending results of evidence analysis and/or additional statements by witnesses. Any other information will be added to this report as a supplemental report. In addition, the sighting location has been identified as a place of interest for further exploration and investigation.

END OF REPORT

This method of report writing has served me well over the years. I like separating portions of the investigation out with headers, and the titles explain what part of the investigation it covers; interviews, scene investigations, etc. You may have also noticed that I like to break up my text with paragraph breaks. One of my pet peeves is when I'm trying to find something in a report hidden in a wall of text. If the text is broken up into short paragraphs, it makes it much easier to come back and review later and find a particular passage.

Typically, after I complete a report, I do not alter it. However, if new information or evidence comes into the case, I like to complete a supplemental report. Doing this preserves the original report but gives an option for adding new information to the case. A supplementary report can be an additional document related to the original report, such as a lab report, weather report, or other documents. Further, if another investigator conducts an interview or other investigation related to the report, you can add that investigator's supplemental report to your case file.

You may find this type of report writing intimidating. It can be time-consuming, and good note-taking is imperative for detailed reports. Even if this type of report does not work for you, I suggest you at least write a synopsis or summarizing report for your investigations. During my time as an investigator, I never regretted documenting too much.

CHAPTER 20
CASE MANAGEMENT

This section covers how to organize your cases into a case file. Some of you may think these steps would be pretty obvious, rendering this section redundant. However, I assure you that there are fledging and veteran investigators out there who will benefit from this information.

I like to organize my case files by year. I suggest on whatever computer or device you plan on storing your files on, you label a file folder "Case Files" or something similar. In that file, you have your year files. There are a lot of file folders within file folders with this method of organization, but I believe it is the most effective. Keeping the individual reports and evidence stored in the proper case file folder can make reviewing the files, and by extension, your life, so much easier.

So, inside your file folder for the year 2022, you will have your file folder labeled 22-003. This is the sighting at Rainbow Lake we wrote the report in the last chapter. Inside your 22-003 folder, you will have your report, any supplemental reports, and more file folders. I like to use a separate file folder for audio and a separate file folder for photos and video. When creating these files, I like to name them something like; 22-003 Audio and 22-003 Photos and Video. I like to do it this way in the event any of the folders accidentally get moved or inadvertently dragged out of their intended case file; it is easy to find where they belong.

If you like, you can create separate folders inside your audio and video folders and sort your digital evidence by the place it originated. For example, in your audio folder, you can have a folder labeled, "Audio from Campsite" and another labeled, "Interview with Witness John D." This is beneficial if you have multiple files from one incident. If you had to stop and start your recorder multiple times during an interview, you might have several separate files that are technically the same. I like to put these in their own folder and keep them separate and organized from other audio files.

Depending on how busy you are, you may have to hold cases in a pending file until you are done with them and ready to move them into their permanent home folder. Starting a report and working on it a section at a time is a good way to establish a workflow. You don't necessarily have to close a case every day, but working on it every day will get you closer and closer to completing it. Maybe you have a day off from your day job, and the weather is too bad to hit the woods? Spend the day working on pending reports.

I suggest keeping an external memory drive available to back up your files. Then, upgrade the unit every few years and keep the old one around as a redundant backup. As the saying goes, "two is one, one is none."

Ultimately, the decision is yours as to how you want to handle the organization of your files. You can be as particular as you want. Compiling a case file combines hours of investigation, audio files, video files, and other evidence into a streamlined, linear narrative that makes sense to you and any outside observers. It creates order from chaos and presents that order to those who need to see it so that they can understand it.

CHAPTER 21
INVESTIGATION TIPS

This chapter is a collection of tips that I learned over the years. They are in no particular order and may or may not apply to your particular style of investigation.

Investigations are not linear. I often had young officers come up to me and ask how to work a sex assault case or how to work a fraud case. Every single time the answer is, "It depends." I wish there was an outline of steps that I could give to every officer on how to work a particular case. The truth is, when we receive cases to investigate, we rarely get the whole story initially. There is no Step A, then Step B, then Step C. Sometimes, you may start with Step C, then go to Step F, only to finally come around to Step A. Bigfoot investigations can be very similar. Be patient, follow the information, and document everything.

You will see an acronym in many detective offices across the country, GOYAAKOD. It stands for, "Get off your ass and knock-on doors." Be proactive. If you hear about sightings in an area, go knock on some doors, pass out some business cards with your contact info, and hand out stickers with your organization's logo. Talk to camp hosts, be an investigator, get out there and investigate!

You can use your camera for much more than taking pictures of scenes and evidence. See a flyer or poster you want to remember? Take a picture of it. Don't want to take someone's last business card? Take a picture of it. Do you want to remember the serial number of a new item you purchased and don't have time to write it down now? Take a picture of it. Do you want to remember the name of a campground, road, or trail and not have your pen and notebook handy? Take a picture of it. Remember, digital film is cheap.

Almost all states have some form of online criminal or court record archive. Some are free, and some cost money. Familiarize yourself with what is available in your area. Learn how to do a basic background search on someone. Before going into the woods with anyone, you should find out what kind of criminal history they have. Never meet with anyone by yourself who has any charge or conviction for a violent crime. Instead, take a friend, or two, to meet with anyone who has a questionable background. Further, beware any witness who has ever been charged with a crime of dishonesty like filing a false police report or perjury. Take anything received from a witness with those types of criminal charges with a liberal dose of skepticism.

The amount of information available online is incredible. I'm not even talking about social media. I'm talking about public records. With online ancestry databases, property records, obituaries, wedding announcements, historical societies, newspaper articles, and a plethora of other resources, the information you can find with a few keystrokes about anyone is amazing, bordering on frightening. As a result, finding the owner of a remote property is now easier than ever.

Forget to write down the date or time something occurred? Check your photos or when the audio file was created. As long as your clock and calendar on your device are accurate, most image and audio files have a date and timestamp that can help you narrow down or pinpoint when a photo was taken or when an audio file was created.

Have a small piece of evidence that needs to be collected but ran out of evidence bags? Pick the item up in your gloved hand and peel the glove off, encasing the item in the glove as you turn it inside out. This will create a small temporary bag you can carry it in without exposing it to contamination. Just be sure you place the glove away from any trash or other items that need to be thrown away and transfer it to a proper evidence bag as soon as you can. I collected many cigarette butts from scenes over the years using this method.

Social media is a great tool. You do not necessarily have to participate, but having an account on some platforms can provide good information. A few years ago, a gentleman was making the rounds in some Bigfoot groups on social media, posting pictures he took of an alleged Bigfoot. A quick glance at his profile revealed that a few weeks prior, he had posted pics of himself posing in his brand new ghillie suit, which looked remarkably like the Bigfoot he had recently captured on film.

Online search engines are your friend. I suggest plugging in combinations of names, dates, places, and other relevant info into any major search engine and seeing what pops up. The hardest part of searching for things online can often be figuring out what combination of words and names can get you the results you are looking for. For example, searching for "Bigfoot at Rainbow Lake" may not get you anything useful, but searching for "Creature at Rainbow Lake" may give you precisely what you are looking for. Same with names of people. Marriages, adoptions, divorces, and other events can lead to name changes and missed or ignored search results.

Create a case worksheet to help keep track of what you have done. It can be a sheet in your notebook or a file on your computer. Document the dates and times of phone calls, even if you did not contact anyone. Leave a voicemail? Put it on your worksheet. Drop your evidence off at the post office to send it to the lab? Put it on your worksheet. This can help you keep track of what you have and have not done. For example, maybe you could never reach a particular person of interest in an investigation. The case worksheet can be a helpful document to keep track of such things. It can be as simple as a form with case number, date/time, and activity columns.

Play the "what if" game. Try to form a plan of action for different scenarios that may play out during your investigations. Realistically, no one will ever be ready for everything. Still, if you plan for likely events and scenarios, you will be more suited to handle the unexpected ones.

As stated earlier, eyewitness testimony is arguably the bread and butter of Bigfoot investigation. However, eyewitness testimony is also some of the most unreliable evidence out there. I cannot count the number of times I spoke to witnesses of a hit and run accident or something similar, and they would all give me a different color for the suspect vehicle. Some would say it was blue, some would say it was green, while another might say yellow. Eventually, when we would track down the vehicle, we would find that one of the witnesses was right, one was close, and one was either lying or color blind. If witnessing something as mundane as a traffic accident can result in such wildly different observations of the same event, think what an incredible paradigm shaking event, like a Bigfoot sighting, can result in. Remember that particularly traumatic events can and do affect memory. In essence, there are two sides to every statement. What the witness remembers seeing, and what the witness actually saw.

When dealing with historical records such as property records or similar, write on the back where and when you got the record when you get a hard copy of a document. This will help you remember the source if you need to go back and look for more. When you begin doing historical research, libraries and archives can all start to blend together. Documenting where you received something can save you a ton of time later.

If possible, always talk to witnesses separately and out of earshot of one another. Hearing another person's observations of the same event can cause people to change their statements to agree with what the other person said.

A box of fresh pastries and a couple of business cards can open more doors than all the phone calls in the world. Some treats and cards dropped off at fish and wildlife offices, dispatch centers, and campground manager's offices can get you more info than any freedom of information act request.

Do not be afraid to ask for advice. Sometimes, sitting down and talking over a case with someone can help you see the next step you need to take. Sharing information with other investigators can help too. Be mindful not to share privileged info like witness names and addresses, but going over a sighting with another investigator can often be a huge help.

Take breaks when you need them. When I was a young rookie police officer, I ate, slept, and breathed police stuff. I read police magazines, watched police shows, and read novels about police investigations. A few years later, I was at the point where I came home, took off my uniform, and did not think about police or law enforcement until it was time to go back to work. I avoided anything that made me even think of the job.

Similar things happen to Bigfoot investigators as well. When I first started looking into the Bigfoot phenomenon, I was a voracious reader of anything Bigfoot and a devout listener to a dozen or so Bigfoot podcasts. Nowadays, I listen to two or three Bigfoot podcasts on a semi-regular basis and only read a Bigfoot book if it is highly recommended by my friends. Burnout is real, and it happens to everyone. Pace yourself.

CHAPTER 22
PARTING WORDS

Being an investigator can be the most rewarding, yet stressful position you can ever be in. The responsibility to collect and present information in an unbiased and factual manner is not for everyone. When you are investigating crimes, you literally hold the lives of those you investigate in your hands. The findings and conclusion of your investigation can affect their lives forever.

Many might say the investigation into the Bigfoot phenomenon, while important, is not as serious as life and death. I disagree. Instead of being responsible for one person's life or reputation, in Bigfoot investigation, you are possibly responsible for the lives of an entire species. Furthermore, proving the existence of Bigfoot to the world could very well start a chain reaction of events that could ensure the preservation of the species that, for all we know, could be near or past the point of no return.

Whatever your motivations for looking for this elusive creature may be. I wish you luck. Because, like any seasoned investigator, I have an almost unquenchable thirst for mysteries and their answers. It is time this mystery gets solved. I often think about the amount of evidence out there, the track casts, the witness statements, the audio and video recordings, and I think about the number of times I have arrested someone and seen them convicted of a crime with much less evidence than there is for our elusive forest giant.

If you have not noticed by now, I love stories. My favorite thing, is to sit around a room full of military and law enforcement veterans and listen to, and tell, stories. One of the instructors at my academy said something along the lines of, "There is a reason why there are so many cop shows on television, and there aren't any shows out there about Joe Blow, the auto parts store clerk." He was right. Those damn few of us that have been in the military and law enforcement have lived through adventures that most people only see on television. Some of those stories, while true, are too unbelievable for tv!

I decided to end this book with this story. It is not a happy story. However, it has a specific purpose. It is a true story. I have done my best to omit details because of the sensitive nature of the story. However, I feel that it still maintains the purpose and gravitas I intended for it to have with these omissions.

Some years ago, in some small town, in some flyover state, a little girl, not yet a teenager, went missing. Her family sounded the alarm almost immediately after she did not return home when she was supposed to. The little town banded together in the way only a small town can. People were going door to door, leaving work early, and forming search parties.

The town was small and only had a few police officers available. One of the officers had a young daughter about the same age as the missing girl. He found this case particularly troubling. So he put on his uniform and came in on his day off to help search.

Not long into the search, a man who was one of the volunteer searchers, found the little girl's body near a lake. Everyone's worst fears were confirmed. She had been murdered. The town was in shock, and the little girl's parents were overwhelmed with grief.

Some local law enforcement and the fire department had voiced suspicion about the man who had found the girl's body. Reportedly, he had not really searched and had seemingly walked right to the spot where the body was found. So, while many of the townspeople and local news outlets celebrated the man for finding the girl's body, local law enforcement was casting a suspicious eye on him.

Our officer, who had come in on his day off, was especially troubled by the discovery of the young girl's body. Not only was it a tragic loss for the entire town, he could not help but think of how close the young girl was to his own daughter's age and how he would feel if placed in the role of the young, murdered girl's parents.

The “hero" was asked to come back to the station and was promptly taken into a room, given some coffee and doughnuts, and told to sit down. Thinking he was about to be debriefed or his statement taken for finding the missing girl, he had no idea he was now the only suspect the police had.

While detectives began to question the "hero" who had found the girl's body, our young officer promptly went to work. Getting a search warrant for the "hero's" car and having it towed to the station.

By this point, it was well past nightfall. It was dark, cold, and raining. Being a small police department, the officer did not have the luxury of having a garage or other indoor area to put the vehicle in to search it. Being hours away from the nearest large city meant not having access to a crime scene unit that could do the job for him. No one was coming. It was up to him. The officer had already been working for several hours. He was cold, wet, hungry, and missing his family. Still, he searched.

He did not just search; he obsessed over every square inch of that car. He was on his knees, looking over the vehicle's floorboard with a flashlight and magnifying glass, wet gravel and grit from the pavement digging into his knees. He searched.

The "hero" denied any involvement in the young girl's disappearance and untimely end. Instead, he argued that the police were jealous that he had found the girl when they could not. Obviously, they did not know what they were doing, and he could do a better job at finding the real killer.

In the rain, outside in the dark, our officer found something. Inside the "hero's" car, he found hair belonging to the little murdered girl. Hair that proved the little girl had been in the vehicle belonging to the man who had found her body hours before, who claimed to be a “hero” for finding her. Hair that eventually led to the confession, charges, and a conviction for the man that kidnapped, murdered, and then beamed as he soaked up attention from news and local citizens after pretending to find her body, where he had left it.

Out in that wet, dark parking lot, that officer found the most crucial piece of evidence in that case, hours past his scheduled return home, hours past his last hot meal, hours past the last contact with his family. He put his family and his own needs on hold while searching in miserable conditions. He brought a murderer to justice, brought a family closure, and removed a monster from society.

He was tested, and he chose to sacrifice, to put in the work, to be great. Someday, somehow, you will be tested; please, when your time comes, be great.

SPECIAL ACKNOWLEDGMENTS

This book would not exist without the training and opportunities provided by the Saint Paul Island Department of Public Safety and the Homer Police Department.

Made in the USA
Monee, IL
28 September 2026

64410354R00115